Advanced Mathematics

Personal Study Notes:

Part I

By

Mohamed F. El-Hewie

BRIEF TABLE OF CONTENTS

TABLE OF CONTENTS

CHAPTER 1: GRADIENT, DIVERGENCE AND CURL

CHAPTER 2: GAUSS'S DIVERGENCE THEOREM AND STOKES' THEOREM

CHAPTER 6: MATRICES **231**

PREFACE

The thought of publishing my personal study notes, that covered fifteen years of undergraduate, graduate, and postdoctoral learning and research, dates to the year 1969. The only difference between my forty year old thought and the present is the phenomenal invention and growth of the Internet and computer software. Then, our copy machines were run by ammonia deposits on chemically treated paper. The stencil tissues and engravers were the common tools for small size publication. None of those means of reproduction was economically viable during my years as student with total dependency on scholarships and family support.

In 1986, after settling in Colorado Springs, Colorado, my elder sister volunteered to ship all my belongings from Egypt to the USA. A half-ton wooden box traveled from Alexandria, Egypt, to Houston, Texas, carrying all my handwritten notes and college textbooks. Then, the custom clearance, brokerage, and transportation between Texas and Colorado ran under $500. The kindness of my sister for risking the enormous troubles in dealing with the Egyptian bureaucratic exporting agencies was paralleled with the similar kindness of the America truck driver who had to drive his 18-wheeler through a residential apartment complex yet to find out that his truck was not equipped with a lift that could lower the half-ton container on the ground of the apartment complex. The driver allowed me an extra day to lease and ready a forklift from a local rental shop. Driving a forklift on the streets of Colorado Springs was one of many adventures I enjoyed in such beautiful Rocky Mountains' town.

By 1986, I already gathered more modern and elegant books in mathematics. But, my personal notes embodied a process of engraving the mathematical thinking into my neurons. My cursive writings and graphic illustrations during the interactive learning with lecturers and independent home searching and preparation are vividly recorded into my forty year old notes. Those include the theoretical subject matter, the examples that followed, and the challenging problems for exercising. I strove to keep the notes as complete as they developed during the learning process. My graphic approach to the interpretation of mathematically abstract concepts illustrates the mental development of my learning. In this phase, the refurbished notes are arranged into linkable topics with bitmapped snips intermingled with searchable text. I hope to publish the remaining notes on other pure and applied mathematics in the manner that could aid the learner in grasping the crux of the subject matter in strategically graphic as well as logical approach. Many of those mathematical concepts give rebirth to newer ones as the learner indulges into the philosophy of the matter with greater focus.

Mohamed F. El-Hewie
Woodland Park, New Jersey February 10, 2011

CHAPTER 1: GRADIENT, DIVERGENCE AND CURL

1. Scalar and Vector fields

If to each point $\underline{r}$ (x, y, z) of a region R in space, there corresponds a scalar function ϕ (x, y, z) ,i.e. $\phi(\underline{r})$, then ϕ is called a scalar function of position or a scalar point function . The complete set of values for ϕ is called the field of values for ϕ defined throughout the region R, or briefly the scalar field ϕ defined throughout R.

If ϕ,has only one value at each point of R, it said to be uniform or single valued, and when all the values are independent of the time t, the field is said to be stationary or a steady field .

For example, the temperature T at any point of the earth's surface and the function ϕ (x, y, z) = cos xcosh y e^{-z} define two scalar fields which are evidently single valued; the latter is steady .

The definitions of vector point fuctions and vector fields are obvious. The velocity of each particle of a mass of a moving fluid or the electric intensity $\underline{E} = \frac{e}{r^3}\underline{r}$ due to a point charge e at the origin are examples of single valued vector fields.

It is usual to imagine in the region R where the scalar function ϕ is defined, surfaces on which ϕ has a constant value, i,e. surfaces whose equation is $\phi(\underline{r})$ = constant. These surfaces are called the level surfaces of ϕ or iso -ϕ. It is evident that no two such surfonces corresponding to different values of ϕ can interesect, because that would mean that ϕ will have two differe-nt values at the pointe of intersection which is contrary to the assumption that the function is single valued . What we call isothermal surfaces and equipotential surfaces are the level surfaces when ϕ is the temperature and potential respectively.

1.1. Definitions of scalar and vector fields

16 – 12 – 1971

Vector Field Theory

Definition . A scalar is a quantity which has a magnitude but no direction for ex. length - area - volume Potential.

. A vector is a quantity which has both magnitude & direction. displacement - velocity - force elect. intensity.

A scalar field is a field such that associated with every pt of the field there is a scalar quantity. e.g. The temp. distribution throughout a solid.

A vector field is a field such that associated with every pt of the field there is a vector quantity. e.g. the velocity distribution in a stream of water.

2. Continuity

A scalar or vector point fuction is said to be continuous at the point P when the difference between its value at a neighbouring point Q and its value at P can be made as small as one pleases by selecting Q sufficiently close to P. In mathematical terms, the point functuin V(r) is sais to be continuous at P(r) if for each positive number ε, we can find another δ such that

$$|V(x+\delta x, y+\delta y, z+\delta z) - V(x,y,z)| < \varepsilon$$

whenever $\delta x < \delta$, $\delta y < \delta$, $\delta z < \delta$. briefly , this can be written in the form $|V(\underline{r} + \delta \underline{r}) - V(\underline{r})| < \varepsilon$ when ever $|\delta \underline{r}| < \delta$. This is equivalent to saying that $V(\underline{r})$ is continous at $\underline{r}$ if

$$\lim_{\substack{\delta x \to 0 \\ \delta y \to 0 \\ \delta z \to 0}} V(x+\delta x, y+\delta y, z+\delta z) = V(x, y, z)$$

or

$$\lim_{\delta r \to 0} V(\underline{r} + \delta r) = V(\underline{r})$$

For vector point functions, the above definition leads to the fact that the function is continuous at any point, if its scalar components in three directions are continuous at the same point.

Since almost all scalar and vector fields which appear in applied mathematics and physics are single valued and continuous, we will always assume that the fields we deal with are of this type.

2. Partial derivatives

The partial derivative of a scalar or vector point function $V(\underline{r})$ with respect to x is defined as

$$\frac{\partial V}{\partial x} = \lim_{\delta x \to 0} \frac{V(x+\delta x, y, z) - V(x,y,z)}{\delta x}$$

Similarly

$$\frac{\partial V}{\partial y} = \lim_{\delta y \to 0} \frac{V(x,y,+\delta y,z) - V(x,y,z)}{\delta y}$$

$$\frac{\partial V}{\partial z} = \lim_{\delta z \to 0} \frac{V(x,y,z+\delta z) - V(x,y,z)}{\delta z}$$

are the partial derivatives of Vwith respect to y and z respectively $\frac{\partial V}{\partial x}$, $\frac{\partial V}{\partial y}$, $\frac{\partial V}{\partial z}$ give the rate of change of V or the directiooal derivatives of V along the x, y, z directions, respectively. If these limits exist, then Vis said to be differentiable.

Higher derivatives can be defined as follows

$$\frac{\partial^2 V}{\partial x^2} = \frac{\partial}{\partial x}\left(\frac{\partial V}{\partial x}\right), \quad \frac{\partial^2 V}{\partial x \partial y} = \frac{\partial}{\partial x}\left(\frac{\partial V}{\partial y}\right)$$

$$\frac{\partial^3 V}{\partial x \partial y^2} = \frac{\partial}{\partial x}\left(\frac{\partial^2 V}{\partial y^2}\right.$$

If $\frac{\partial^2 V}{\partial x \partial y}$ and $\frac{\partial^2 V}{\partial y \partial x}$ are both continuous, then they are equal.

3. Increments in point functions

Let $V(\underline{r})$ be defined and differentiable at each point $\underline{r}$ in a certain region . If $\underline{r}$ changes by $\delta \underline{r}$, then the corresponding change δV in V is

$$\delta V = V(x + \delta x,\ y + \delta y,\ z + \delta z) - V(x,y,z)$$

$$= \frac{\partial V}{\partial x}\delta x + \frac{\partial V}{\partial y}\delta y + \frac{\partial V}{\partial z}\delta z + \ldots$$

$$= \left(\delta x \frac{\partial}{\partial x} + \delta y \frac{\partial}{\partial y} + \delta z \frac{\partial}{\partial z}\right) V + \ldots$$

Since $\delta \underline{r} = \underline{i}\,\delta x + \underline{j}\,\delta y + \underline{k}\,\delta z$ and if we define the operator ∇ by

$$= \underline{i}\frac{\partial}{\partial x} + \underline{j}\frac{\partial}{\partial y} + \underline{k}\frac{\partial}{\partial z}$$

then

$$\delta V = (\delta \underline{r} \cdot \nabla) V + \ldots.$$

to the first order of small quantities.

Using the notation of differentials, we can write

$$dV = (d\underline{r} \cdot \nabla) V$$

∇ is called the vector differential operator and read del or nable. It is an operator that has the properties of vectors which do not contradict its being an operator, e.g. $\nabla V \neq V \nabla$

5. Directional derivatives

Consider a scalar or vector function $V(\underline{r})$ which is defined and continuous in a region R. Let it have the values $V = V(\underline{r})$ and $V + \delta V = V(\underline{r} + \delta \underline{r})$ at two neighbouring points $P(\underline{r})$ and $Q(\underline{r} + \delta \underline{r})$. If PQ is in the direction of the unit vector $\underline{a}$ and $PQ = \delta s$, then $\delta \underline{r} = \underline{a}\, \delta s$ and $\delta V = \delta s\, (\underline{a} \cdot \nabla) V + \ldots$ The directional derivative or rate of change of V along the direction $\underline{a}$ is

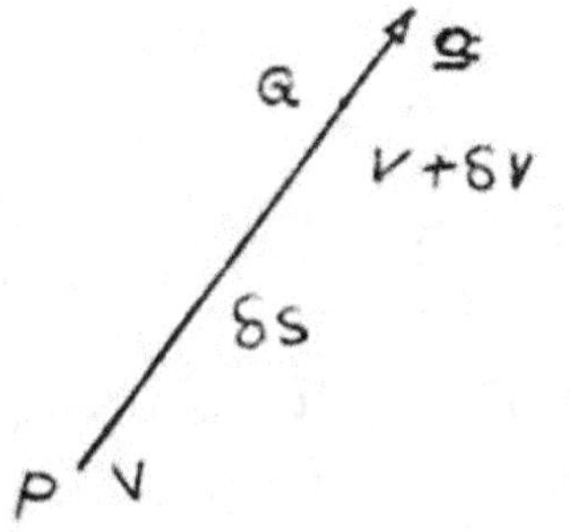

$$\frac{\partial V}{\partial s} = \lim_{\delta s \to 0} \frac{V(r + \delta r) - V(r)}{\delta s} = (\underline{a} \cdot \nabla) V$$

6. The gradient of a scalar function

i) Let the scalar function ϕ be defined and differentiable at every point $\underline{r}$ in a certain region of space (i.e. ϕ defines a differentiable scalar field), then the gradient of ϕ – written grad ϕ or $\nabla \phi$ – is defined to be

$$\nabla \phi = \left(\underline{i} \frac{\partial}{\partial x} + \underline{j} \frac{\partial}{\partial y} + \underline{k} \frac{\partial}{\partial z} \right) \phi$$

$$= \underline{i} \frac{\partial \phi}{\partial x} + \underline{j} \frac{\partial \phi}{\partial y} + \underline{k} \frac{\partial \phi}{\partial z}$$

Therefore, $\nabla \phi$ is a vector point function whose components are $\frac{\partial \phi}{\partial x}$, $\frac{\partial \phi}{\partial y}$, $\frac{\partial \phi}{\partial z}$, it is along the normal to the level surface ϕ = constant in the direction ϕ increases

Definition

A vector field whose vector point function $\underline{V}$ is the gradient of some scalar function ϕ (i.e. $\underline{V} = \nabla\phi$) is called a scalar potential field; ϕ is its scalar potential .

6.1. Direction of gradient vector:

ii) We have seen that the directional derivative of ϕ along the direction $\underline{a}$ at P is $\frac{\partial \phi}{\partial s} = \lim_{\delta s \to 0} \frac{\delta \phi}{\delta s}$

$$= (\underline{a} \cdot \nabla)\phi = \underline{a} \cdot (\nabla\phi)$$

where δs is the length PQ in the dirextion $\underline{a}$

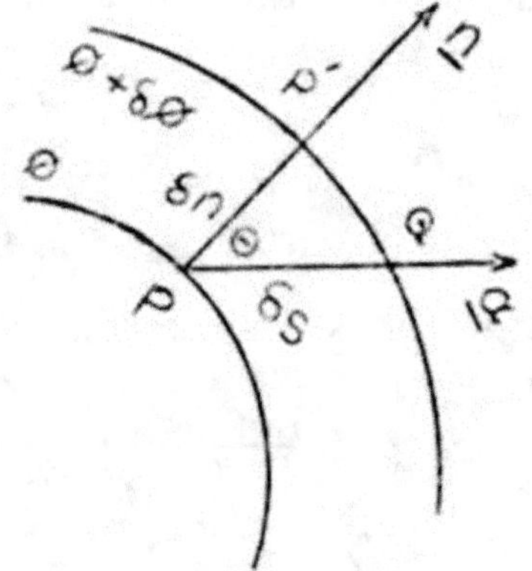

Consider the level surfaces at P , Q with values ϕ , $\phi + \delta\phi$ respectively . Lest $\underline{n}$ be the unit vector normal to the first surface at P in the direction in which ϕ increases ; this with intersect the level surface at Q at a point P, say . If pp'= n, then the directional derivative of ϕ in the direction $\underline{n}$ is

$$\frac{\partial \phi}{\partial n} = \lim_{\delta n \to 0} \frac{\delta \phi}{\delta n} = \underline{n} \cdot \nabla \phi.$$

$$= \sqrt{\left(\frac{\partial \phi}{\partial x}\right)^2 + \left(\frac{\partial \phi}{\partial y}\right)^2 + \left(\frac{\partial \phi}{\partial z}\right)^2}$$

since $\underline{n} = \left(i \frac{\partial \phi}{\partial x} + \underline{j} \frac{\partial \phi}{\partial y} + \underline{k} \frac{\partial \phi}{\partial z}\right) \Big/ \sqrt{\left(\frac{\partial \phi}{\partial x}\right)^2 + \left(\frac{\partial \phi}{\partial y}\right)^2 + \left(\frac{\partial \phi}{\partial z}\right)^2}$

Therefore, from i) $\qquad \nabla \phi = \underline{n} \frac{\partial \phi}{\partial n}$

This shows that, although we used some fixed rectangular axes of coordinates, the vector function $\nabla \phi$ is independent of our choice of these axes . Also

$$\frac{\partial \phi}{\partial s} = \underline{a} \cdot \nabla \phi = (\underline{a} \cdot \underline{n}) \frac{\partial \phi}{\partial n} = \cos \theta \frac{\partial \phi}{\partial n}$$

where θ is the angle between $\underline{a}$ and $\underline{n}$. Hence the greatest directional derivative of a scalar function ϕ at any point, is that along the normal to the level surface of ϕ at that point.

6.2. Practical Application of the gradient

6.2.1. Electrostatic field

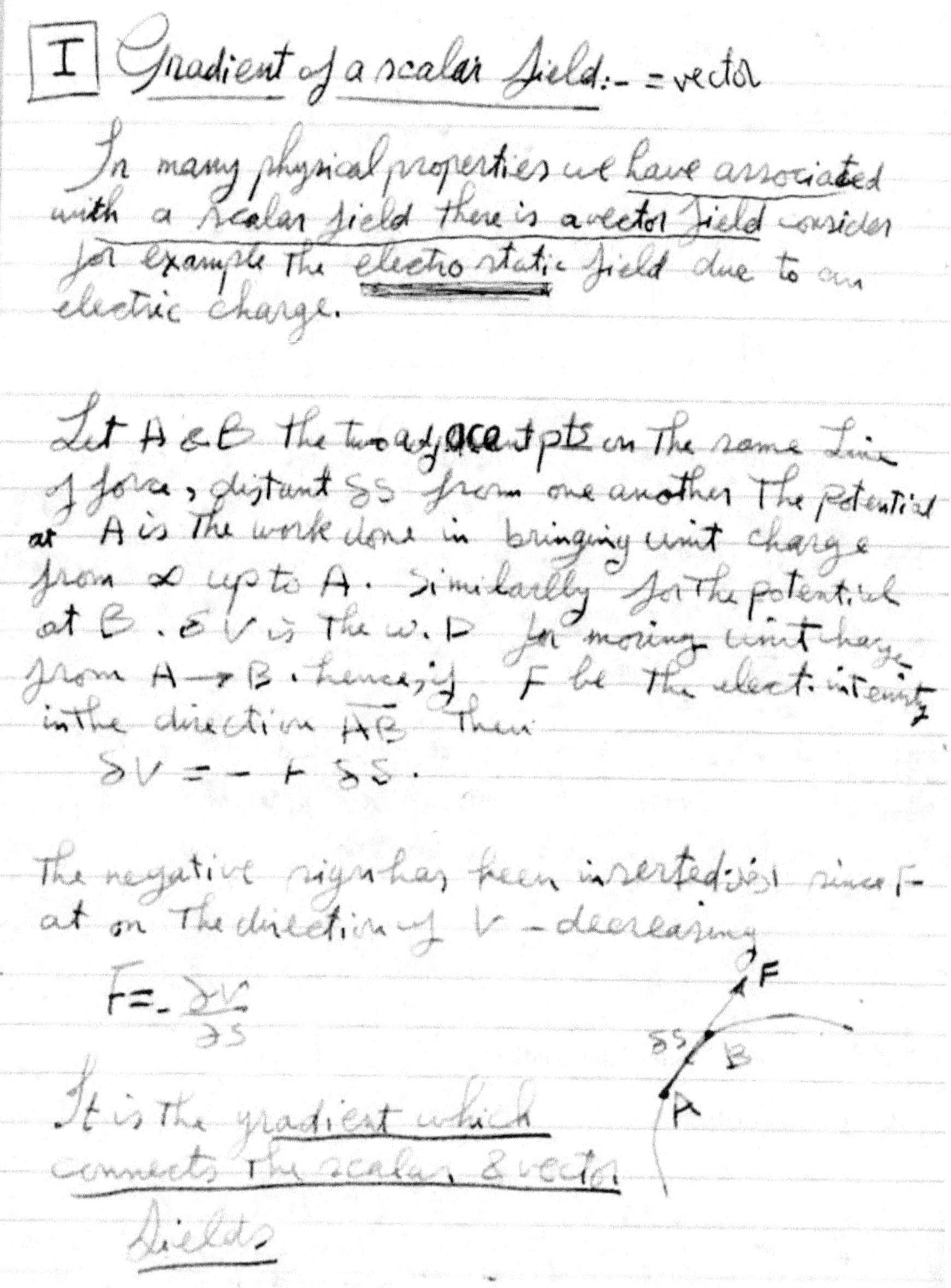

I Gradient of a scalar field:- = vector

In many physical properties we have associated with a scalar field there is a vector field consider for example the electro static field due to an electric charge.

Let A & B the two adjacent pts on the same line of force, distant δs from one another The potential at A is the work done in bringing unit charge from ∞ up to A. Similarlly for the potential at B. δV is the W.D for moving unit charge from A → B. hence, if F be the elect. intensity in the direction $\overline{AB}$ Then

$\delta V = - F \, \delta s.$

The negative sign has been inserted: as since F- act on The direction of V - decreasing

$F = -\frac{\partial V}{\partial s}$

It is the gradient which connects the scalar & vector fields

6.2.2. Equipotential surface or level surface

Let $\phi(x, y, z)$ be a scalar pt assumed single-valued continious and possessing continious space derivatives

- We now defined what is known as a level surface or an equipotential surface. This is a:

A surface on which ϕ has the same value of all pts on the surfaces.

- We next defind a del or Nabla operator ∇

$$\nabla \equiv \frac{\partial}{\partial x}\bar{i} + \frac{\partial}{\partial y}\bar{j} + \frac{\partial}{\partial z}\bar{k}$$

This is a directive differential operator which may operate either on a scalar fn or on a vector fn. • We now defind the gradient of the scalar fn ϕ as follows.

$$\text{grad}\,\phi = \nabla\phi = \left(\frac{\partial}{\partial x}\bar{i} + \frac{\partial}{\partial y}\bar{j} + \frac{\partial}{\partial z}\bar{k}\right)\phi$$

$$= \frac{\partial \phi}{\partial x}\cdot\bar{i} + \frac{\partial \phi}{\partial y}\bar{j} + \frac{\partial \phi}{\partial z}\cdot\bar{k}$$

Which is a vector whose scalar component are

$\frac{\partial \phi}{\partial x}$, $\frac{\partial \phi}{\partial y}$, $\frac{\partial \phi}{\partial z}$

6.2.3. Interpretation of the magnitude and direction of the gradient of a scalar field

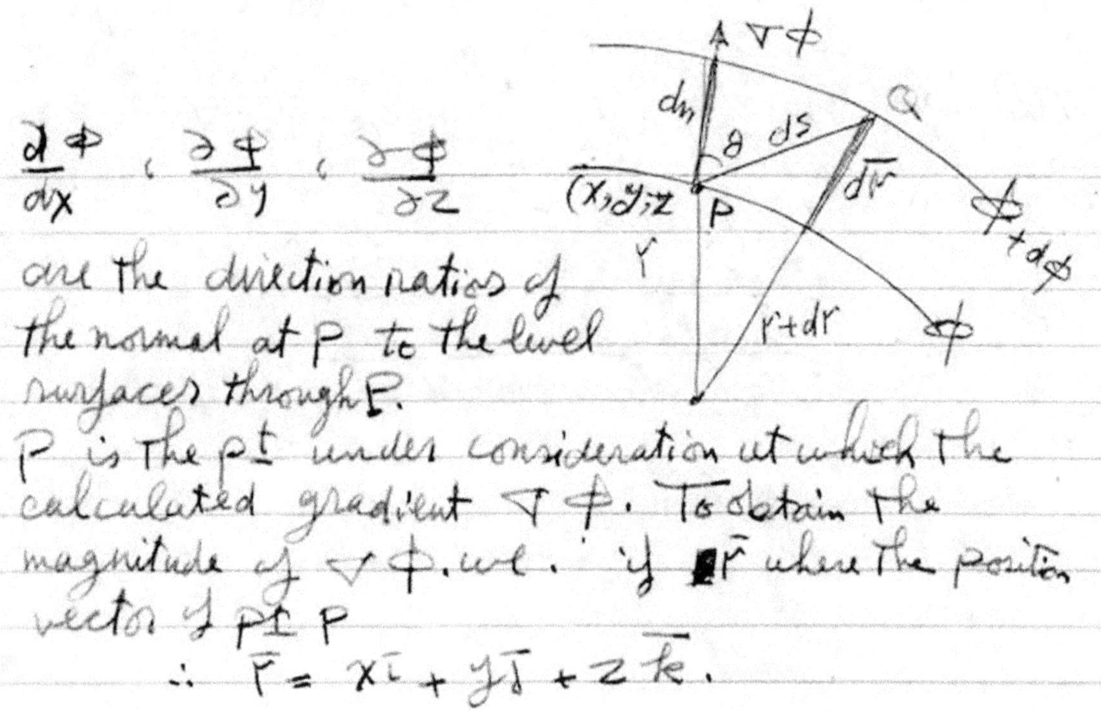

$\frac{\partial \phi}{\partial x}$, $\frac{\partial \phi}{\partial y}$, $\frac{\partial \phi}{\partial z}$ are the direction ratios of the normal at P to the level surfaces through P.

P is the pt under consideration at which the calculated gradient $\nabla\phi$. To obtain the magnitude of $\nabla\phi$, we. if $\bar{r}$ where the position vector of pt P

$$\therefore\ \bar{r} = x\bar{i} + y\bar{j} + z\bar{k}.$$

Hence

$$\nabla \phi = \frac{\partial \phi}{\partial x}\bar{i} + \frac{\partial \phi}{\partial y}\bar{j} + \frac{\partial \phi}{\partial z}\bar{k}.$$

& $\overline{dr} = dx\, i + dy\, \bar{j} + dz\, \bar{k}$.

<u>multiplying</u>

$$\nabla \phi \cdot \overline{dr} = \frac{\partial \phi}{\partial x}dx + \frac{\partial \phi}{\partial y}dy + \frac{\partial \phi}{\partial z}dz = d\phi$$

$$\therefore |\nabla \phi|\, ds \cos\theta = d\phi$$

$$|\nabla \phi|\, dn = d\phi \qquad |\nabla \phi| = \frac{\partial \phi}{\partial n}$$

Hence the magnitude of vector $\nabla \phi$ represents the rate of change of ϕ in the direction of normal at P to the level surface through P. This represents the maximum rate of change of ϕ

6.2.4. Example of the gradient derivation

Now consider the rate of change of ϕ in the direction PQ this is given by $\frac{\partial\phi}{\partial s}$. since $dn = ds\cos\theta$

$$\therefore \frac{\partial\phi}{\partial s} = \frac{\partial\phi}{\partial n}\cos\theta = |\nabla\phi|\cos\theta.$$

Here the directional derivatives of ϕ in the direction PQ is equal to the scalar component of $\nabla\phi$ in that direction.

ex. calculate the directional derivatives of scalar fn $\phi(x,y,z) = xy^2 + yz^3$ at the pt $(2,-1,1)$ in the direction of vector $(\bar{i} + 2\bar{j} + 2\bar{k})$

Solⁿ

$$\phi = xy^2 + yz^3$$

$$\frac{\partial\phi}{\partial x} = y^2 \;(\quad \frac{\partial\phi}{\partial y} = 2xy + z^3 \;, \quad \frac{\partial\phi}{\partial z} = 3yz^2$$

$$\frac{\partial\phi}{\partial n} = \nabla\phi = \frac{\partial\phi}{\partial x}\bar{i} + \frac{\partial\phi}{\partial y}\bar{j} + \frac{\partial\phi}{\partial z}\bar{k}$$

$$\nabla\phi = y^2\bar{i} + (2xy + z^3)\bar{j} + 3yz^2\bar{k}$$

substitute by pt $(2,-1,1)$ ⟶ ①

$$\nabla\phi = \bar{i} - 3\bar{j} - 3\bar{k}$$

scalar product.

$|\nabla\phi| \times 1 \times \cos\phi$

$\nabla\phi$

$1 + 2\bar{j} + 2\bar{k}$

are direction ratios.

$$\nabla\phi = i - 3j - 3\bar{k}$$

$$ds = i + 2j + 2k$$

$$\alpha i + \beta j + \gamma k$$

$$= \frac{\alpha i + \beta j + \gamma k}{\sqrt{\alpha^2 + \beta^2 + \gamma^2}} = \frac{i + 2j + 2k}{\sqrt{1+4+4}}$$

$$= m\bar{i} + n\bar{j} + lk = \frac{1}{3}i + \frac{2}{3}j + \frac{2}{3}k$$

$$\therefore |\nabla\phi| \cos\theta = \frac{1}{3} - 3\times\frac{2}{3} - 3\times\frac{2}{3}$$

$$= -\frac{11}{3}$$

we obtained the rate of change of ϕ in the direction of the vector $\bar{i} + 2\bar{j} + 2\bar{k}$ by forming the scalar product of $\nabla\phi$ and unit vector.

7. Line integrals

Let V be scalar or vector point function defined and continuous in a region R. Let C be a curve in R joining two points O and A. Divide C into a large number n of elemental arcs; let PQ be one of these elements, its length being δs. The limit

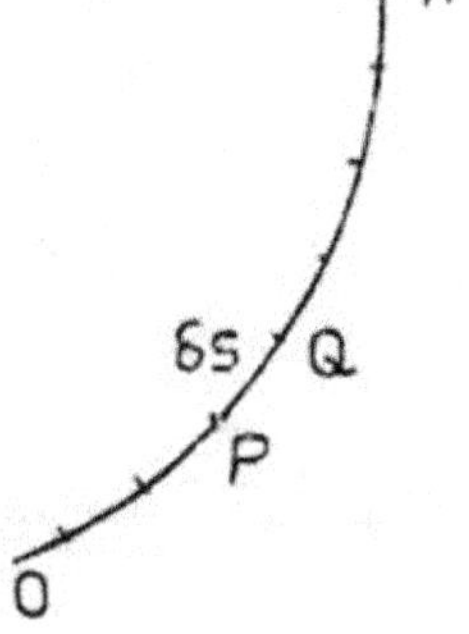

$$\lim_{n\to\infty} \sum V\, ds \quad (\text{or } \lim_{\delta s \to 0} \sum V ds),$$

is called the line integral of V along the curve C and written as $\int_C V \; ds$.

If C is a simple closed curve, i.e. a curve which does not intersect itself anywhere, then the integral round C is often denoted by $\oint_C V \; ds$.

Two particular forms appear frequently in applied mathematics. These are

$$\int_C \underline{F} \cdot \underline{T} \, ds \quad \text{and} \int_C \underline{F} \wedge \underline{T} \; ds,$$

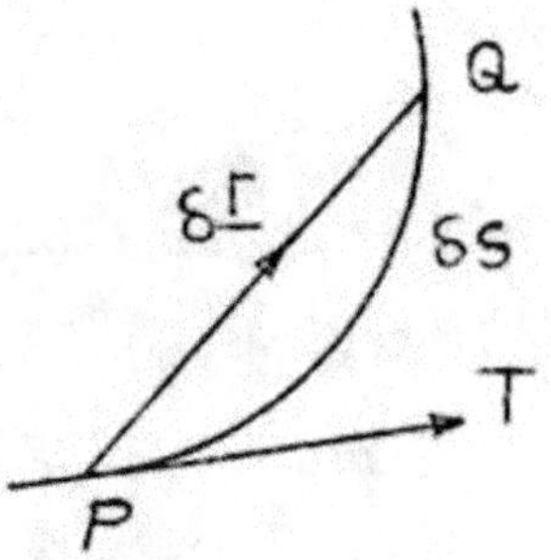

Where $\underline{T}$ is the unit vector along the tangent to C at P. Since

$$\overrightarrow{PQ} = \delta \underline{r} = \underline{T} \; \delta s$$

to the frist order of small quantities, the above two integrals take also the forms

$$\int_C \underline{F} \cdot d\underline{r} \quad \text{and} \quad \int_C \underline{F} \wedge d\underline{r}$$

In the first integral we are, in fact, integrating the tangential component of $\underline{F}$.
For example, when $\underline{F}$ is the force, $\int_C \underline{F} \cdot d\underline{r}$ gives the work done by $\underline{F}$ when the point of application moves along C. In fluid mechanics, the quantity $\oint_C \underline{q} \cdot d\underline{r}$, where $\underline{q}$ ($\underline{r}$) is the velocity of the fluid particle at $\underline{r}$, is an important one. It is known as the circulation round C.

7.1. Application of linear integrals

Linear integrals are useful in simulating the work done in electrostatic fields, flow of fluid and energy along change in gradient.

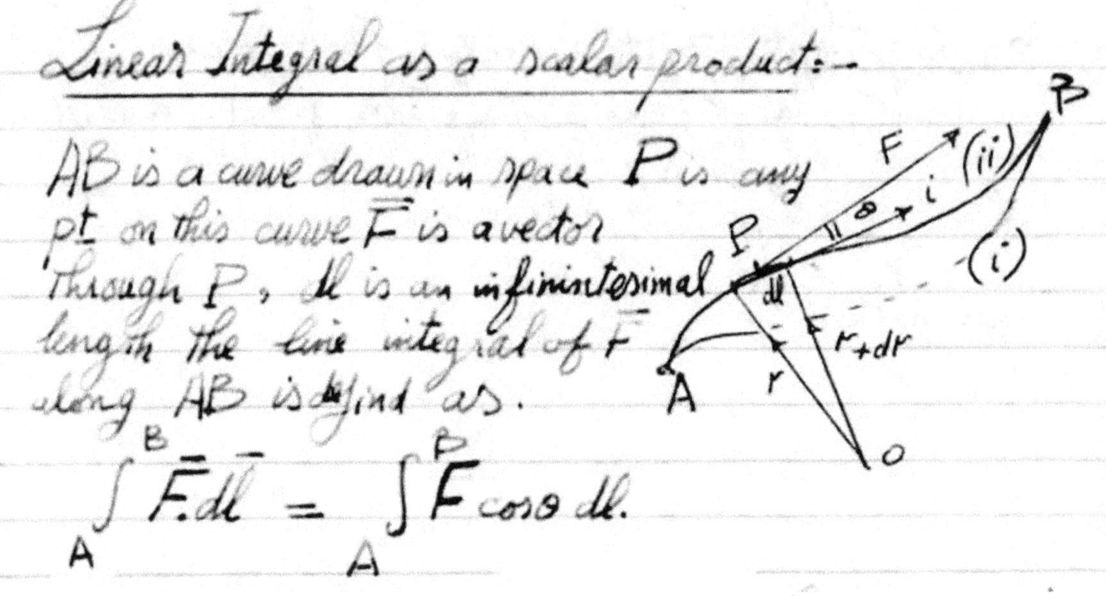

Linear Integral as a scalar product:-

AB is a curve drawn in space P is any pt on this curve $\bar{F}$ is a vector through P, dl is an infinintesimal length the line integral of F along AB is defind as.

$$\int_A^B \bar{F}.\bar{dl} = \int_A^B F \cos\theta \, dl.$$

This means that we multiply the scalar component of $\bar{F}$ along the tangent at P multiplied by the element of arc & integrate.

Line integral is defined as above can be interpreted physically as follows:-

1 - If $\bar{F}$ represents a force acting on particle then the line integral represents the work done a moving particle from A $\xrightarrow{to}$ B.
2. If $\bar{F}$ represents electric intensity at P in an electrostatic field then line integral represents potentials diff^ce^ bet^n^ A & B.
3. If $\bar{F}$ represents the velocity of fluid at P then the Line integral of $\bar{F}$ along closed path gives the circulation arround the path

8. Conservative Fields

A vector field $\underline{V}$ defined in a region R is said to be concervative if $\int_{OP} \underline{V} \cdot d\underline{r}$ is independent of the curve in R joining O to P , or equivalently if $\oint_C \underline{V} \cdot d\underline{r} = 0$ for every closed curve C in R

Consodcr the field $\underline{V} = \nabla \phi$

$$\int_{OP} \nabla \phi \cdot d\underline{r} = \int_{OP} d\phi. = \phi_P - \phi_O$$

i.e. the integral depends only on the points P and Q . Also, for any closed curve C,

$$\oint_C \nabla \phi \cdot d\underline{r} = 0,$$

as ϕ is single valued.

Therefore every scalar potential field is conservative

Conversely , if $\oint_C \underline{V} \cdot dr = 0$ for every dlosed curve C in R , then $\underline{V}$ is the gradient of some scalar function ϕ . For consider any closed curve OSPQ , then

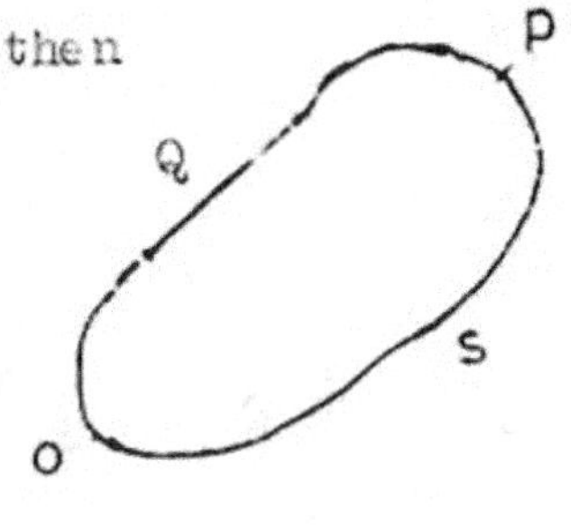

$$\oint_{OSPQO} V.dr = 0$$

i.e. $\int_{OSP} \underline{V}.d\underline{r} = - \int_{OQP} \underline{V}.d\underline{r}$

$$= \int_O^P \underline{V}.d\underline{r} \text{ along any curve in R joining O to P.}$$

Let O be a fixed point and P a variable one. Then $\int_O^P \underline{V} \cdot d\underline{r}$ is a scalar function of position of P which we may denote by ϕ ; i.e.

$$\phi = \int_O^P \underline{V} . d\underline{r}$$

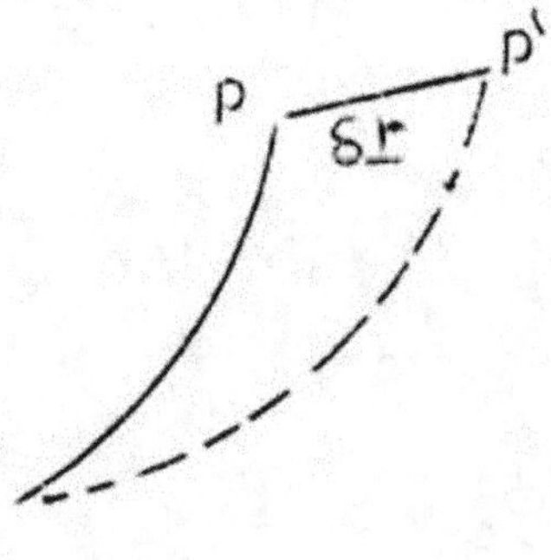

For a neighbouring point P' such that $PP' = \delta \underline{r}$

$$\phi + \delta\phi = \int_O^{P'} \underline{V}.d\underline{r}$$

$$\therefore \delta\phi = \int_P^{P'} \underline{V}.\, d\underline{r} = \underline{V}.\ \delta\underline{r}$$

But $\delta\phi = \delta\underline{r}.\ \nabla\phi$. Hence

$(\underline{V} - \nabla\phi).\delta\underline{r} = 0$ for all values of $\delta\underline{r}$

$$\therefore \underline{V} = \nabla\phi .$$

Therefore, every conservative field $\underline{V}$ is scalar potential ; its potential function is $\phi(r) = \int_O^{\underline{r}} \underline{V}.\ d\underline{r}$.
This integral is calculated along any curve joining some standard position O to $\underline{r}$

8.1. Application on electrostatic field

Now let us consider the important but particular case when $\bar{F} = \nabla \phi$

i.e: There is a scalar potential fn ϕ with gradient of the vector $\bar{F}$

e.g: electric field $d\bar{E} = -\frac{dV}{dx}$ $\quad E = -\nabla V$

$$\bar{F} = \nabla \phi = \frac{\partial \phi}{\partial x}\bar{i} + \frac{\partial \phi}{\partial y}\bar{J} + \frac{\partial \phi}{\partial z}k.$$

$$\bar{dl} = d\bar{r} = dx\,i + dy\,J + dz\,k.$$

$$\therefore \quad \vec{r} = x\bar{i} + y\bar{J} + z\,k.$$

$$\therefore \quad \bar{F}\cdot\bar{dl} = \nabla\phi\cdot d\bar{r} = \frac{\partial \phi}{\partial x}dx + \frac{\partial \phi}{\partial y}dy + \frac{\partial \phi}{\partial z}dz$$

$$= d\phi$$

$$\therefore \quad \int_A^B \bar{F}\cdot\bar{dl} = \int_A^B d\phi = [\phi]_A^B = \phi_B - \phi_A.$$

i.e the integral in this case <u>is not depend</u> on the <u>path</u> But depend <u>only</u> on the <u>ends</u>. points of the line integral of $\bar{F}$ path(i) is equal to the line integral of $\bar{F}$ along path(ii) is equal to......

Hence from this it follow:- if we integrate from
A to B along path (i) & the Back from
B to A along path (ii) the sum = 0
since the direction which the path takes change
the sign of integral in the reverse path.

Hence the field of F = the line integral of F along
a closed path equal to zero, such a field is called:

"A scalar potential field or lamellar field"

It is called scalar potential due to the property of a
scalar potential fn ϕ with grad F.
It is called lamellar (consist of layer or thin plates) by
level surfaces or equipotential surfaces on each of
which ϕ is const.

The properties of Lamellar field are :- ($F = \nabla\phi$)

1. The integral from A to B is independent on the path, only depends on the ends.

2. The line integral alonge a closed path is zero.

3. Since we have shown that curl grad ϕ : $\nabla \times \nabla \phi = 0$ such a scalar potential field is irrotational field. i.e the irrotational field is at the same time a scalar potential field. i.e exact with grad F.

8.2. Gradient of a single valued differentiable function

i) For a single valued differentiable function f (r)

$$\nabla f(r) = \mathbf{i}\frac{\partial f}{\partial x} + \mathbf{j}\frac{\partial f}{\partial y} \quad \mathbf{k}\frac{\partial f}{\partial z}$$

$$= \frac{df}{dr}\left(\mathbf{i}\frac{\partial r}{\partial x} + \mathbf{j}\frac{\partial r}{\partial y} + \mathbf{k}\frac{\partial r}{\partial z}\right)$$

Since $r^2 = x^2 + y^2 + z^2$,

then $\frac{\partial r}{\partial x} = \frac{x}{r}$, $\frac{\partial r}{\partial y} = \frac{r}{y}$, $\frac{\partial r}{\partial z} = \frac{r}{z}$ and

$$\nabla f(r) = \frac{df}{dr}\left(\mathbf{i}\frac{x}{r} + \mathbf{j}\frac{y}{r} + \mathbf{k}\frac{z}{r}\right)$$

$$= \frac{\mathbf{r}}{r}\frac{df}{dr}$$

Alternatively, the level surfaces of $f(r)$ are the concentric spheres r = constant, and the unit vector normal to any of these spheres at any point is along the radius, i. e. is along the position vector $\underline{r}$ of this point.

$$\nabla f(r) = \underline{n} \frac{df}{dn} = \frac{\underline{r}}{r} \frac{df}{dr}$$

In particular

$$\nabla r^m = m\, r^{m-2}\, \underline{r} .$$

8.3. Gradient of a constant vector

ii) For a constant vector $\underline{a} = \underline{i}\, a_x + \underline{j}\, a_y + \underline{k}\, a_z$,

$$\underline{a}.\underline{r} = a_x x + a_y y + a_z z$$

and

$$\nabla(\underline{a}.\underline{r}) = \underline{i} \frac{\partial}{\partial x} (\underline{a}.\underline{r}) + \underline{j} \frac{\partial}{\partial y} (\underline{a}.\underline{r}) + \underline{k} \frac{\partial}{\partial z} (\underline{a}.\underline{r})$$

$$= \underline{i}\, a_x + \underline{j}\, a_y + \underline{k}\, a_z = \underline{a} .$$

9. The divergence of a vector

Let $\underline{V} = \underline{i}V_x + \underline{j}V_y + \underline{k}\, V_z$ be defined and differentiable at each point $\underline{r}$ in a region of space.

The divergence of $\underline{V}$, written div $\underline{V}$ or $\nabla.\underline{V}$, is defined by

$$\nabla . \underline{V} = (\underline{i} \frac{\partial}{\partial x} + \underline{j} \frac{\partial}{\partial y} + \underline{k} \frac{\partial}{\partial z}) . (\underline{i}\, V_x + \underline{j}\, V_y + \underline{k}\, V_z)$$

$$= \frac{\partial V_x}{\partial x} + \frac{\partial V_y}{\partial y} + \frac{\partial V_z}{\partial z} .$$

9.1. Interpretation of divergence of a vector

II. The divergence of a vector. = scalar

This is Def :- as the scalar product of nabla operator & the vector it self.

i.e $\quad div\ \bar{F} = \nabla \cdot \bar{F} = \left(\frac{\partial}{\partial x}\bar{i} + \frac{\partial}{\partial y}\bar{j} + \frac{\partial}{\partial z}\bar{k}\right) \cdot (F_x i + F_y j + F_z k)$

$$\nabla \cdot \bar{F} = \frac{\partial F_x}{\partial x} + \frac{\partial F_y}{\partial y} + \frac{\partial F_z}{\partial z}$$ which is a

scalar quantity. to obtain a tangible of meaning let us suppose that the velocity vector $\bar{v}$ of the fluid particle its scalar, components in the x - y - z - direction are V_x, V_y, V_z.

Quantity of fluid entering the element Through the surface ① in time Δt

$$= \delta y\, \delta z \left[V_x - \frac{\partial V_x}{\partial x}\frac{\partial x}{2}\right] \partial t \longrightarrow ①$$

P is The centre of the element.

as $y = f(x)$

$$\delta y = \frac{dy}{dx}\,\delta x.$$

The quantity of fluid leaving the element from face ② in the same interval of time δt is

$$= \delta y\,\delta z\left(V_x + \frac{\partial V_x}{\partial x}\cdot\frac{\partial x}{2}\right)\partial t \longrightarrow ②$$

The net out flow from ① & ②

$$\frac{\partial V_x}{\partial x}\,\delta y\,\delta z\,\delta x\,\delta t.$$

We have other similar expressions hence the total divergence of fluid from the element in time δt

$$\left(\frac{\partial V_x}{\partial x} + \frac{\partial V_y}{\partial y} + \frac{\partial V_z}{\partial z}\right)\delta x\,\delta y\,\delta z\,\delta t.$$

Hence the out flow per unit time is

$$\frac{\partial V_x}{\partial x} + \frac{\partial V_y}{\partial y} + \frac{\partial V_z}{\partial z} \quad \text{i.e. } \nabla.\overline{V}.$$

if the fluid is incompressible & there are no sources or sinks within the element then there can be neither gain or loss in general volume element and hence there is The Net out flow

$$\nabla.\overline{V} = \frac{\partial V_x}{\partial x} + \frac{\partial V_y}{\partial y} + \frac{\partial V_z}{\partial z} = 0 \quad \text{Cond}^n \text{ of continuity}$$

Given what is known at the eqn of continuity. if the divergence zero at all pts of a vector field the field is said to be (Solenoidal) ... in such a field the flux convergance on an element is exactly blanced by an equal flux divergence ... from the element.

ex. show that the vector field given by.

$\bar{F} = y\cos(2x+y^2)\,\bar{i} - \cos(2x+y^2)\,\bar{j} + x^2y\,\bar{k}$

is solenoidal.

$\nabla \cdot \bar{F} = -2y\sin(2x+y^2) + 2y\sin(2x+y^2) + 0 = 0$

10. The curl of a vector

The curl or rotation of $\underline{V}$, written curl $\underline{V}$, rot $\underline{V}$ or $\nabla \wedge \underline{V}$, is defined by

$$V = \left(\underline{i}\frac{\partial}{\partial x} + \underline{j}\frac{\partial}{\partial y} + \underline{k}\frac{\partial}{\partial z}\right) \wedge \left(\underline{i}V_x + \underline{j}V_y + \underline{k}V_z\right)$$

$$= \underline{i}\left(\frac{\partial V_z}{\partial y} - \frac{\partial V_y}{\partial z}\right) + \underline{j}\left(\frac{\partial V_x}{\partial z} - \frac{\partial V_z}{\partial x}\right) + \underline{k}\left(\frac{\partial V_y}{\partial x} - \frac{\partial V_x}{\partial y}\right).$$

$$= \begin{vmatrix} \underline{i} & \underline{j} & \underline{k} \\ \frac{\partial}{\partial x} & \frac{\partial}{\partial y} & \frac{\partial}{\partial z} \\ V_x & V_y & V_z \end{vmatrix}$$

We will prove later that $\nabla \cdot \underline{V}$ and $\nabla \wedge \underline{V}$ are invariant with respect to the choice of rectangular axes. It is easy to prove that

$$\nabla \cdot \underline{r}\, f(r) = r \frac{df}{dr} + 3f \,,$$

$$\nabla \wedge \underline{r}\, f(r) = 0 \,,$$

and for a constant vector $\underline{a}$

$$\nabla \cdot (\underline{a} \wedge \underline{r}) = 0 \,,$$

$$\nabla \wedge (\underline{a} \wedge \underline{r}) = 2\, \underline{a}$$

10.1. Interpretation of curl of a vector

III Curl of vector.

This is defined as the vector product of Nubla operator and the vector itself.

$$\text{Curl}\, \bar{F} = \nabla \times \bar{F} = \begin{vmatrix} i & j & \bar{k} \\ \frac{\partial}{\partial x} & \frac{\partial}{\partial y} & \frac{\partial}{\partial z} \\ F_x & F_y & F_z \end{vmatrix}$$

$$\text{curl } \bar{F} = \left(\frac{\partial F_z}{\partial y} - \frac{\partial F_y}{\partial z}\right)\bar{i} + \left(\frac{\partial F_x}{\partial z} - \frac{\partial F_z}{\partial x}\right)\bar{j} + \left(\frac{\partial F_y}{\partial x} - \frac{\partial F_x}{\partial y}\right)\bar{k}$$

to obtain a tangible meaning for the curle of a vector let us consider the case of a rigid body rotating with an angular velocity ω

about an axis OA where O is a point fixed in the body [illegible].

Hence the pt P has a tangential velocity $\bar{\omega} \times \bar{r}$. now suppose that the body as a whole has a linear velocity of transilatation $\bar{V_0}$ hence the velocity $\bar{v}$ of the pt P is:

$$\bar{v} = \bar{V_0} + \bar{\omega} \times \bar{r}.$$

$$\text{curl } \bar{v} = \text{curl } \bar{V_0} + \text{curl } \bar{\omega} \times \bar{r}.$$

[illegible]

$$\text{curl } \bar{V_0} = 0 \quad \text{translation only}$$

Since $\bar{V_0}$ is the same for all pts x, y, z of the body.

$$\bar{\omega} \times \bar{r} = \begin{vmatrix} \bar{i} & \bar{j} & \bar{k} \\ \omega_x & \omega_y & \omega_z \\ x & y & z \end{vmatrix}$$

$$= (\omega_y z - \omega_z y)\bar{i} + (\omega_z x - \omega_x z)\bar{j} + (\omega_x y - \omega_y x)\bar{k}$$

$$\text{curl } \bar{\omega} \times \bar{r} = \begin{vmatrix} \bar{i} & \bar{j} & \bar{k} \\ \frac{\partial}{\partial x} & \frac{\partial}{\partial y} & \frac{\partial}{\partial z} \\ (\omega_y z - \omega_z y) & (\omega_z x - \omega_x z) & (\omega_x y - \omega_y x) \end{vmatrix}$$

$$i = \frac{\partial}{\partial y}(\omega_x y - \omega_y x) - \frac{\partial}{\partial z}(\omega_z x - \omega_x z) = \omega_x + \omega_x$$

$$\bar{j} = \frac{\partial}{\partial z}(\omega_y z - \omega_z y) - \frac{\partial}{\partial x}(\omega_x y - \omega_y x) = \omega_y + \omega_y$$

$$\bar{k} = \frac{\partial}{\partial x}(\omega_z x - \omega_x z) - \frac{\partial}{\partial y}(\omega_y z - \omega_z y) = \omega_z + \omega_z$$

$\therefore \operatorname{curl} \bar{V} = 2\bar{\omega}$.

i.e. The curl of the total velocity vector is twice the angular velocity vector. if in a vector field, ~~the field is said to be irrotational otherwise its rotational~~ The curl vanishes at pts of the field, then the field is said to be irrotational, otherwise it is rotational.

condition of irrotationality

$$\boxed{\nabla \times F = 0}$$

10.2. Alternative definition of the curl of a vector

6-1- 1971.

Alternative Defination of the curl of a vector.

Let P is any pt in a vector field, surround P by a plane infinitesimal area, Boundaried by a a closed curve C. The line Integral.

$\int_C \bar{F}.\bar{dl}$ has a certain value.

•P

C

Now rotate the area about P for every position of the area the line integral

$\int_C \bar{F}.\bar{dl}$ has a certain value & for some positions the line integral is maximum. if we divide the max. line integral of F by the area & take the limit when the area shrinks to a point we get the magnitude of curl F & the plane of area is then normal to curl F.

$$\lim_{area \to 0} \frac{Max \int_C \bar{F}.\bar{dl}}{area} = |\nabla \times F|$$

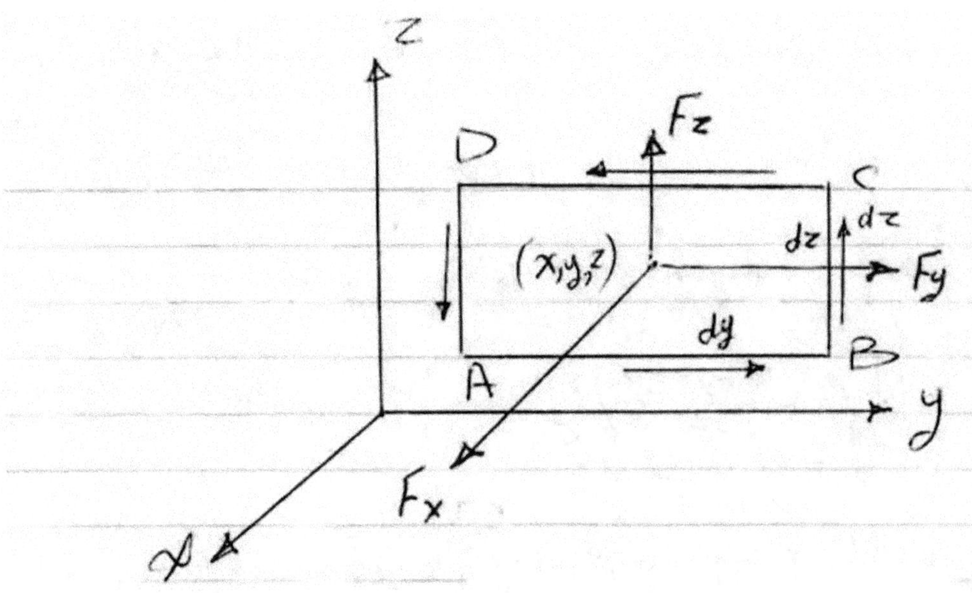
z
Fz
D
C
dz
dz
(x,y,z)
Fy
dy
A
B
y
Fx
x

1. Scalar component of F along AB $= F_y - \frac{\partial F_y}{\partial z} \cdot \frac{dz}{2}$

2. " " " " BC $= F_z + \frac{\partial F_z}{\partial y} \cdot \frac{dy}{2}$

3. " " " " " DC $= F_y + \frac{\partial F_y}{\partial z} \cdot \frac{dz}{2}$

4. " " " " DA $= F_z - \frac{\partial F_z}{\partial y} \frac{dy}{2}$.

The Line integral of F along ABCD is

$$\left.\begin{aligned} &\left(F_y - \frac{\partial F_y}{\partial z} \frac{dz}{2}\right) dy \\ + &\left(F_z + \frac{\partial F_z}{\partial y} \cdot \frac{dy}{2}\right) dz \\ + &(-)\left(F_y + \frac{\partial F_y}{\partial z} \frac{\partial z}{2}\right) dy \\ + &(-1)\left(F_z - \frac{\partial F_z}{\partial y} \frac{dy}{2}\right) dz \end{aligned}\right\} = \left(\frac{\partial F_z}{\partial y} - \frac{\partial F_y}{\partial z}\right) dy\,dz$$

dividing by area $dy\,dz$ of the element we get

$\frac{\partial F_z}{\partial y} - \frac{\partial F_y}{\partial z}$ which is the scalar component of curl F in the direction of x-axis i.e in the direction normal to the area.

$$\begin{vmatrix} i & j & k \\ \frac{\partial}{\partial x} & \frac{\partial}{\partial y} & \frac{\partial}{\partial z} \\ F_x & F_y & F_z \end{vmatrix}$$

$\nabla \times \bar{F}$

normal

θ

Line integral of $E = |\nabla \times \bar{F}| \cos\theta$
area

The line integral is max when $\theta = 0$ that is the plane of the area.

ex. A vector field is given by

$$\bar{F} = (x^2 + xy^2)\,i + (y^2 + x^2 y)\,j$$

show that the field is irrotational & evalute the integral from (0, 1) to (1, 2) & show that this equal the potential difference between this two pts.

Soln

$$\nabla \times \bar{F} = \begin{vmatrix} i & j & k \\ \frac{\partial}{\partial x} & \frac{\partial}{\partial y} & \frac{\partial}{\partial z} \\ x^2 + xy^2 & y^2 + x^2 y & 0 \end{vmatrix}$$

$$= \frac{\partial}{\partial z}(y^2 + x^2 y) i + \frac{\partial}{\partial z}(x^2 + xy^2) j + \frac{\partial}{\partial x}(y^2 + x^2 y) k - \frac{\partial}{\partial y}(x^2 + xy^2) k$$

$$= (2yx - 2xy) k = 0$$

That is hence field is irrotational as $\nabla \times \bar{F} = 0$

hence — (1)

$$\bar{F} = \nabla \phi$$

$$F_x i + F_y j + F_z k = \left(\frac{\partial \phi}{\partial x}\right) i + \left(\frac{\partial \phi}{\partial y}\right) \bar{j} + \frac{\partial \phi}{\partial z} k$$

$$\frac{\partial \phi}{\partial x} = x^2 + xy^2 \quad \text{— (1)}$$

$$\frac{\partial \phi}{\partial y} = y^2 + x^2 y \quad \text{— (2)}$$

$$\frac{\partial \phi}{\partial z} = 0$$

integrate the 1st eqn w.r.t x we get

$$\phi = \frac{x^3}{3} + \frac{x^2 y^2}{2} + \psi(y)$$

No differentiate w.r.t y we get

$$\frac{\partial \phi}{\partial y} = + x^2 y + \psi'(y) \quad \text{— (3)}$$

from (2), (3)

$$= y^2 + x^2 y$$

$\therefore \psi'(y) = y^2 \quad \therefore \psi(y) = \frac{y^3}{3} + C$ ← لا يؤثر

$\therefore \phi = \frac{x^2}{3} + \frac{x^2 y^2}{2} + \frac{y^3}{3}$.

integration →

$$\int_A^B \bar{F} \cdot dl$$

$$= \int_A^C \bar{F} dl + \int_C^B \bar{F} dl.$$

$$= \int_A^C F_x dx + F_y dy \quad (dy \to 0)$$

$$+ \int_C^B F_x dx + F_y dy \quad (dx \to 0)$$

y

$B\ (1,2)$

$(0,1)$ A

C

$y = 1$, $dy = 0$

$x = 1$, $dx = 0$

x

$$= \int_0^1 (x^2+xy^2)\,dx \quad \underline{\text{with } y=1}$$

$$+ \int_1^2 (y^2+x^2y)\,dy \quad \underline{\text{with } x=1}$$

$$= \int_0^1 (x^2+x)\,dx + \int_1^2 (y^2+y)\,dy = \frac{14}{3} \rightarrow (2)$$

this should be equal to potential diff^ce

$$= [\phi]_{0,1}^{1,2} = \left[\frac{x^3}{3} + \frac{x^2y^2}{2} + \frac{y^3}{3}\right]_{0,1}^{1,2}$$

$$\text{i.e} = \phi_B - \phi_A$$

The vector field $\boxed{\bar{F} = \nabla\left(\frac{k}{r}\right)}$

i.e

$$\phi = \frac{k}{r}$$

this is one of the most important "lamellar" (صفائحي) or scalar potential fields.

$$\bar{F} = \left(\frac{\partial}{\partial x} i + \frac{\partial}{\partial y} J + \frac{\partial}{\partial z} K\right) \frac{k}{r}$$

where $r = \sqrt{x^2+y^2+z^2}$

$$\therefore \bar{F} = \left(\frac{\partial}{\partial x}\bar{i} + \frac{\partial}{\partial y}\bar{j} + \frac{\partial}{\partial z}\bar{k}\right) k(x^2+y^2+z^2)^{-\frac{1}{2}}$$

$$= -\frac{k}{2}(x^2+y^2+z^2)^{-\frac{3}{2}}(2x\bar{i} + 2y\bar{j} + 2z\bar{k})$$

$$= -\frac{k}{r^3}(x\bar{i} + y\bar{j} + z\bar{k})$$

$\bar{F} = -\frac{k}{r^3}\bar{r}$ if $\hat{r}$ is unit vector along the radius vector $\bar{r}$

$\therefore \hat{r} = \frac{\bar{r}}{r}$

$\therefore$ $\boxed{\bar{F} = -\frac{k}{r^2}\hat{r}}$ which is the inverse square law.

$$\bar{\nabla}\cdot\bar{F} = \left(\frac{\partial}{\partial x}\bar{i} + \frac{\partial}{\partial y}\bar{j} + \frac{\partial}{\partial z}\bar{k}\right)\left(-\frac{k}{r^3}[x\bar{i} + y\bar{j} + z\bar{k}]\right)$$

$$= \frac{\partial}{\partial x}\left[\frac{-kx}{(x^2+y^2+z^2)^{\frac{3}{2}}}\right] + \frac{\partial}{\partial y}\left[\frac{-ky}{(x^2+y^2+z^2)^{\frac{3}{2}}}\right] + \frac{\partial}{\partial z}\left[\frac{-kz}{(x^2+y^2+z^2)^{\frac{3}{2}}}\right]$$

$$= \frac{-k}{(x^2+y^2+z^2)^3}\left[(x^2+y^2+z^2)^{\frac{3}{2}} - x\cdot\frac{3}{2}(x^2+y^2+z^2)^{\frac{1}{2}}2x\right.$$

$$+ (x^2+y^2+z^2)^{\frac{3}{2}} - y\,\tfrac{3}{2}(x^2+y^2+z^2)^{\frac{1}{2}}\,2y$$

$$+ (x^2+y^2+z^2)^{\frac{3}{2}} - z\,\tfrac{3}{2}(x^2+y^2+z^2)^{\frac{1}{2}}\,2z$$

$$= -\frac{k}{(x^2+y^2+z^2)^{\frac{3}{2}}}\left[3(x^2+y^2+z^2)^{\frac{3}{2}} - (x^2+y^2+z^2)^{\frac{1}{2}}(3x^2+3y^2+3z^2)\right]$$

$$= 0 \qquad \boxed{\nabla\cdot\bar{F} = 0} \text{ Continiuty}$$

Now $\bar{F} = \nabla\phi$

$\therefore \nabla\cdot\bar{F} = \nabla\cdot(\nabla\phi) = \nabla^2\phi = 0$

That is ϕ satisfys Laplace's eqns hence for such field we have

i) $\phi = \frac{k}{r}$ iii) $\nabla\cdot\bar{F} = 0$

ii) $F = -\frac{k}{r^2}\hat{r}$ vi) $\nabla^2\phi = 0$

If $k = 1$ then $\phi = \frac{1}{r}$ such field occurs in the case of.

i) Ionic charge
ii) Unit magnetic pole.
iii) unit gravitating mass
iv) unit force or sink

Such field are known as Newtonian fields.

11. Solenoidal fields

The selonoidal fields have been defined amidst the discussion on the divergence of a vector. Here, we will recap the previous definition of a solenoidal field.

A vextor field $\underline{V}$ is said to be selenoidal if div $\underline{V} = 0$. The electrostatic field $\underline{E}$ in space if it is free from charge and the velocity field $\underline{q}$ of an incompressible fluid in motion are both solenoidal.

The Field $\nabla \wedge \underline{V}$ is solenoidal if $\underline{V}$ has continuous second order partial derivatives. For

$$\operatorname{div}(\nabla \wedge \underline{V}) = \frac{\partial}{\partial x}\left(\frac{\partial V_z}{\partial y} - \frac{\partial V_y}{\partial z}\right) + \frac{\partial}{\partial y}\left(\frac{\partial V_x}{\partial z} - \frac{\partial V_z}{\partial x}\right) + \frac{\partial}{\partial z}\left(\frac{\partial V_y}{\partial x} - \frac{\partial V_x}{\partial y}\right)$$

Conversely, if $\underline{V}$ is solenoidal, a function $\underline{F}$ can be found such that $\underline{V} = \nabla \wedge \underline{F}$. consider the vector function $\underline{F}$ whose components are

$$F_x = \int V_y \, dz, \quad F_y = -\int V_x \, dz, \quad F_z = 0;$$

the integration is taken along a path parallel to the z-axis

$$= \begin{vmatrix} \underline{i} & \underline{j} & \underline{k} \\ \frac{\partial}{\partial x} & \frac{\partial}{\partial y} & \frac{\partial}{\partial z} \\ F_x & F_y & 0 \end{vmatrix} = \underline{i}\, V_x + \underline{j}\, V_y - \underline{k}\left(\frac{\partial}{\partial x}\int V_x \, dz + \frac{\partial}{\partial y}\int V_y \, dz\right)$$

$$= \underline{i}\, V_x + \underline{j}\, V_y - \underline{k} \int \left(\frac{\partial V_x}{\partial x} + \frac{\partial V_y}{\partial y} \right) dz$$

$$= \underline{i}\, V_x + \underline{j}\, V_y + \underline{k} \int \frac{\partial V_z}{\partial z}\, dz \qquad = \underline{i} V_x + \underline{j}\, V_y + \underline{k}\, V_z$$

Therefore a function $\underline{F}$ is found such that $\underline{V} = \nabla \wedge \underline{F}$

However, this function is still orbitrary to the extent of the addition of the gradient of a scalar function for

$$\text{curl } (\underline{F} + \nabla \phi) = \text{curl } \underline{F};$$

since

$$\text{Curl grad } \phi = \begin{vmatrix} \underline{i} & \underline{j} & \underline{k} \\ \frac{\partial}{\partial x} & \frac{\partial}{\partial y} & \frac{\partial}{\partial z} \\ \frac{\partial \phi}{\partial x} & \frac{\partial \phi}{\partial y} & \frac{\partial \phi}{\partial z} \end{vmatrix} = \underline{i}\left(\frac{\partial^2 \phi}{\partial y \partial z} - \frac{\partial^2 \phi}{\partial z \partial y} \right) + \underline{j} .. + \underline{k} .$$

= 0 when ϕ has continuous second order partial derivatives.

The function $\underline{F}$ is called the vector potential of the field $\underline{V}$ and $\underline{V}$ is sometimes called vector potential field.

12. Irrotational fields

Likewise, irrotational fields have been discussed above amidst the discussion of the curl of a vector. Again, we will recap the definition of irrotational field as follows.

These are vector fields $\underline{V}$ such that $\nabla \wedge \underline{V} = 0$

Let $\underline{V}$ be defined and differentiable in a region R. Applying Stoles's theroem -this will be proved in the next chapter-to any surface S in R we get

$$\oint_C \underline{V} \cdot d\underline{r} = \int_S (\nabla \wedge \underline{V}) \cdot \underline{n}\, ds,$$

where C is the closed curve bounding S.

If the field is irrotational, $\nabla \wedge \underline{V} = 0$ and $\oint_C \underline{V} \cdot d\underline{r} = 0$

for every closed curve in R. Therefore $\underline{V}$ is conservative.
Conversely, if V is conservative, i.e. $\oint_C \underline{V} . d\underline{r}$ vanishes for every closed curve C in P, then

$$\int_S (\nabla \wedge . \underline{V}) \cdot \underline{n}\ ds. = 0$$

for all surfaces S in R. Thus $(\nabla \wedge \underline{V}) \cdot \underline{n}\ \delta S = 0$

for all elements of area δS in R. Hence $\nabla \wedge \underline{V} = 0$

at every point and the field is irrotational.

Therefore, scalar potential fields, conservativefields and irrotational fields are just one class of fields.

13. Expansion formulae

i) Formulae involving sums of functions

$$\nabla \sum \phi_i = \sum \nabla \phi_i ,$$

$$\nabla \cdot \sum \underline{V}_i = \sum \nabla \cdot \underline{V}_i ,$$

$$\nabla \wedge \sum \underline{V}_i = \sum \nabla \wedge \underline{V}_i .$$

ii) Formulae involving products of two functions

The following formula may be found useful in deducing the formulae a to f below. Its proof may be found in any book of vector algebra.

$$\nabla * (uv) = \nabla * (u_o v) + \nabla * (uv_o) ,$$

where * stands for any type of operation for ∇ , and the product between brackets may be ordinary , scalar or vector product according to the type of functions u, v. Thus the order of the factors in each product is in general relevent. The suffix o indicates that the suffixed function is not operated on by ∇ . In each of the two terms on the right hand side, ∇ is to be treated as an ordinary vector in making the suitable changes in the order of terms till ∇ operates on the function without the suffix o.

a) $\text{grad}\,\phi\psi = \phi\,\text{grad}\,\psi + \psi\,\text{grad}\,\phi$

b) $\text{div}\,\phi\underline{V} = \phi\,\text{div}\,\underline{V} + \underline{V}.\,\text{grad}\,\phi$

c) $\text{curl}\,\phi\underline{v} = \phi\,\text{curl}\,\underline{V} + (\text{grad}\,\phi) \wedge \underline{V}$

d) $\text{div}\,(\underline{U} \wedge \underline{V}) = \underline{U}.\text{curl}\,\underline{V} + \underline{V}.\text{curl}\,\underline{U}$

e) $\text{curl}\,(\underline{U} \wedge \underline{V}) = (\underline{V}.\nabla)\,\underline{U} - (\underline{U}.\nabla)\,\underline{V} + \underline{U}\,\text{div}\,\underline{V} - \underline{V}\text{div}\,\underline{U}$

f) $\text{grad}\,(\underline{U}.\underline{V}) = (\underline{V}.\nabla)\,\underline{U} + (\underline{U}.\nabla)\,\underline{V} + \underline{V} \wedge \text{curl}\,\underline{U} + \underline{U} \wedge \text{curl}\,\underline{V}$

14. Second order identities and expressions

In the following we assume that $\phi, \underline{v}$ has continuous second order partial derivatives.

i) $\text{curl grad}\,\phi = \nabla \wedge (\nabla\,\phi) = 0$,

ii) $\text{div curl}\,\underline{v} = \nabla . (\nabla \wedge \underline{V}) = 0.$

iii) $$\text{div grad}\,\phi = \text{div}\left(\underline{i}\frac{\partial\phi}{\partial x} + j\frac{\partial\phi}{\partial y} + k\frac{\partial\phi}{\partial z}\right) = \frac{\partial^2\phi}{\partial x^2} + \frac{\partial^2\phi}{\partial y^2} + \frac{\partial^2\phi}{\partial z^2}.$$

This is usually written in the form $\nabla^2\phi$. The operator

$$\nabla^2 = \frac{\partial^2}{\partial x^2} + \frac{\partial^2}{\partial y^2} + \frac{\partial^2}{\partial z^2}$$

is called Laplace's operator and is unvariant with respect to rectangular axes. It may be applied to a vector functio, provided we interpret the result according to the formula

$$\nabla^2\,\underline{V} = \frac{\partial^2\underline{V}}{\partial x^2} + \frac{\partial^2\underline{V}}{\partial y^2} + \frac{\partial^2\underline{V}}{\partial z^2}$$

iv) curl curl $\underline{V}$ = grad div $\underline{V} - \nabla^2\underline{V}$

It is left to the student to derive this result using cartesian coordinates.

The above four results are very important and should be remembered. This can be easily achieved by treating ∇ formally as an ordinary vector.

15. Summary of vector relationships

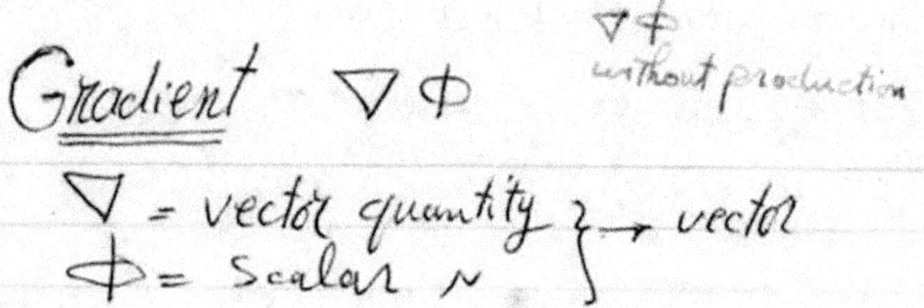

$$P = \phi(x, y, z)$$

$$\nabla = \frac{\partial}{\partial x} i + \frac{\partial}{\partial y} J + \frac{\partial}{\partial z} k$$

$$r = x i + y J + z k.$$

$$\nabla\phi = \frac{\partial\phi}{\partial x} i + \frac{\partial\phi}{\partial y} J + \frac{\partial\phi}{\partial z} k. \quad (1)$$

$$dr = dx\, i + dy\, J + dz\, k \quad (2)$$

$$\nabla\phi \cdot dr = \frac{\partial\phi}{\partial x} dx + \frac{\partial\phi}{\partial y} dy + \frac{\partial\phi}{\partial z} dz.$$ elec. static field.

scalar product $= |\nabla\phi||dr|\cos\theta$

n θ ds dr
$P = \phi(x, y, z)$
r

$\nabla\phi . dr = d\phi$ ⟶ (3)

$\therefore \boxed{\nabla\phi = \frac{d\phi}{dr}}$ ∴ The Gradient of any scalar quantity is the rate of change of that quantity w.r.t displacement.

results

I: $\frac{d\phi}{ds} = \frac{d\phi}{dr}\cos\theta$

If S given by eqn $= x'i + y'J + z'k$.

$$\cos\theta = \frac{x'i + y'j + z'k}{\sqrt{x'^2 + y'^2 + z'^2}}$$

$$\frac{d\phi}{dr} = \frac{\partial\phi}{\partial x}i + \frac{d\phi}{\partial y}J + \frac{d\phi}{\partial z}k$$

$\frac{d\phi}{dr}\cos\theta$ ✓

Divergence $\nabla . F$

$\nabla =$ vector quantity . ⟶ scalar quantity

$F =$ vector quantity

Fluid enters from face (1) $= \left(V_x - \frac{\partial V_x}{\partial x}\frac{dx}{2}\right) dz\,dy\,dt$

F. leaves face (2) $= \left(V_x + \frac{\partial V_x}{\partial x}\frac{dx}{2}\right) dz\,dy\,dt$

Net liquid $= \frac{\partial V_x}{\partial x}\,dz\,dy\,dx\,dt$.

Z, N_z, V_y, y, N_x, (1), (2), (3)

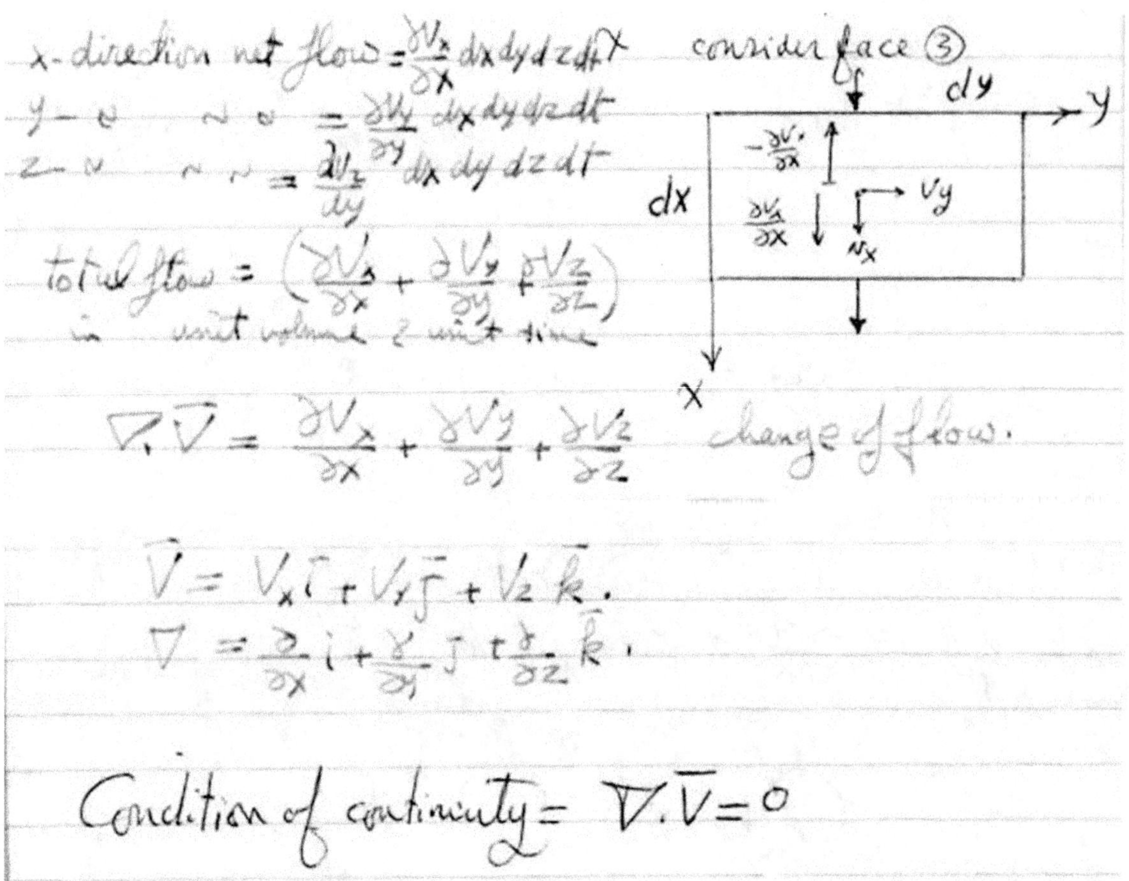

x-direction net flow $= \frac{\partial V_x}{\partial x} dx\,dy\,dz\,dt$

y- " " " $= \frac{\partial V_y}{\partial y} dx\,dy\,dz\,dt$

z- " " " $= \frac{\partial V_z}{\partial y} dx\,dy\,dz\,dt$

total flow in unit volume & unit time $= \left(\frac{\partial V_x}{\partial x} + \frac{\partial V_y}{\partial y} + \frac{\partial V_z}{\partial z}\right)$

consider face ③

$$\nabla \cdot \bar{V} = \frac{\partial V_x}{\partial x} + \frac{\partial V_y}{\partial y} + \frac{\partial V_z}{\partial z}$$ change of flow.

$$\bar{V} = V_x \bar{i} + V_y \bar{j} + V_z \bar{k}.$$

$$\nabla = \frac{\partial}{\partial x} i + \frac{\partial}{\partial y} j + \frac{\partial}{\partial z} k.$$

Condition of continuity $= \nabla \cdot \bar{V} = 0$

Curl. $\nabla \times F$

∇ = vector quantity } vector.
F = vector quantity }

rotation work =

$u\,dx + \left(v + \frac{\partial v}{\partial x}dx\right)dy - \left(u + \frac{\partial u}{\partial y}dy\right)dx$

$+ (-v\,dy) = \left(\frac{\partial v}{\partial x} - \frac{\partial u}{\partial y}\right)dx\,dy$

In unit area rotation in direction $z = \left(\frac{\partial v}{\partial x} - \frac{\partial u}{\partial y}\right)$

in direction $x = \left(\frac{\partial w}{\partial y} - \frac{\partial v}{\partial z}\right)$

" " $y = \left(\frac{\partial u}{\partial z} - \frac{\partial w}{\partial x}\right)$

$$\nabla \times F = \left(\frac{\partial w}{\partial y} - \frac{\partial v}{\partial z}\right)i + \left(\frac{\partial u}{\partial z} - \frac{\partial w}{\partial x}\right)j + \left(\frac{\partial v}{\partial x} - \frac{\partial u}{\partial y}\right)k$$

$$= \begin{vmatrix} i & j & k \\ \frac{\partial}{\partial x} & \frac{\partial}{\partial y} & \frac{\partial}{\partial z} \\ u & v & w \end{vmatrix}$$

$$F = u\bar{i} + v\hat{j} + w\bar{k}$$

$$\nabla = \frac{\partial}{\partial x} i + \frac{\partial}{\partial y} j + \frac{\partial}{\partial z} k.$$

Fundamental vector relation

Gradient:-

I. 1 $$\boxed{\nabla(\phi_1 + \phi_2) = \nabla\phi_1 + \nabla\phi_2}$$

$$\nabla(\phi_1 + \phi_2) = \left(\frac{\partial}{\partial x}\bar{i} + \frac{\partial}{\partial y}j + \frac{\partial}{\partial z}k\right)(\phi_1 + \phi_2)$$

$$= \left(\frac{\partial \phi_1}{\partial x} i + \frac{\partial \phi_1}{\partial y} j + \frac{\partial \phi_1}{\partial z} k\right)$$

$$+ \left(\frac{\partial \phi_2}{\partial x}\bar{i} + \frac{\partial \phi_2}{\partial y} j + \frac{\partial \phi_2}{\partial z} k\right)$$

$$= \nabla\phi_1 + \nabla\phi_2.$$

i.e. $\mathrm{div}(\bar{B} + \bar{A}) = \mathrm{div}\bar{A} + \mathrm{div}\bar{B}$

II Divergence :-

2

$$\boxed{\nabla\cdot(\bar{A}+\bar{B}) = \nabla\cdot\bar{A} + \nabla\cdot\bar{B}}$$

$$\nabla\cdot(\bar{A}+\bar{B}) = \frac{\partial}{\partial x}(\bar{A}_x+\bar{B}_x) + \frac{\partial}{\partial y}(\bar{A}_y+\bar{B}_y) + \frac{\partial}{\partial z}(\bar{A}_z+\bar{B}_z)$$

$$= \left(\frac{\partial}{\partial x}A_x + \frac{\partial A_y}{\partial y} + \frac{\partial A_z}{\partial z}\right) + \left(\frac{\partial B_x}{\partial x} + \frac{\partial B_y}{\partial y} + \frac{\partial B_z}{\partial z}\right)$$

$$= \nabla\cdot\bar{A} + \nabla\cdot\bar{B}$$

Note :-

$$\nabla = \frac{\partial}{\partial x}i + \frac{\partial}{\partial y}j + \frac{\partial}{\partial z}k.$$

$$\bar{A} = \bar{A}_x i + \bar{A}_y j + \bar{A}_z k.$$

$$\bar{B} = B_x i + B_y j + B_z k.$$

$$\bar{A}+\bar{B} = (A_x+B_x)i + (A_y+B_y)j + (A_z+B_z)k$$

i.e. $div\ \phi\bar{A} = \phi\ div\bar{A} + \bar{A}\ grad\ \phi.$

IV

1

$$\boxed{\nabla\cdot(\phi\bar{A}) = \phi(\nabla\cdot\bar{A}) + \bar{A}(\nabla\phi)}$$

$$\nabla\cdot(\phi\bar{A}) = \phi\ div.\bar{A} + \bar{A}\cdot Grad\ \phi.$$

$$= \phi\ \nabla\cdot\bar{A} + \bar{A}\cdot\nabla\phi$$

scalar scalar vector vector.

الأول في الثاني + المشتق للأول × الثاني

Note: $\nabla = \frac{\partial}{\partial x} i + \frac{\partial}{\partial y} \bar{j} + \frac{\partial}{\partial z} k$

$\Phi = \Phi(x, y, z)$

$\bar{A} = A_x i + A_y \bar{j} + A_z \bar{k}$

$$\therefore \nabla \cdot (\Phi \bar{A}) = \nabla \cdot (\Phi A_x \bar{i} + \Phi A_y J + \Phi A_z \bar{k})$$

$$= \frac{\partial}{\partial x}(\Phi A_x) + \frac{\partial \Phi A_y}{\partial y} + \frac{\partial \Phi A_z}{\partial z}$$

$$= \left(\Phi \frac{\partial A_x}{\partial x} + A_x \frac{\partial \Phi}{\partial x}\right) + \left(\Phi \frac{\partial A_y}{\partial y} + A_y \frac{\partial \Phi}{\partial y}\right)$$

$$+ \left(\Phi \frac{\partial A_z}{\partial z} + A_z \frac{\partial \Phi}{\partial z}\right)$$

$$= \Phi\left(\frac{\partial A_x}{\partial x} + \frac{\partial A_y}{\partial y} + \frac{\partial A_z}{\partial z}\right)$$

$$+ \left(A_x \frac{\partial \Phi}{\partial x} + A_y \frac{\partial \Phi}{\partial y} + A_z \frac{\partial \Phi}{\partial z}\right)$$

$$= \Phi \nabla \cdot \bar{A} + (A_x i + A_y \bar{j} + A_z k) \cdot \left(\frac{\partial \Phi}{\partial x} i + \frac{\partial \Phi}{\partial y} j + \frac{\partial \Phi}{\partial z}\right.$$

$$= \Phi \nabla \cdot \bar{A} + \bar{A} \cdot \nabla \Phi$$

$$= \Phi \, div \bar{A} + \bar{A} \, Grad \, \Phi$$

ب.(∇ x أ) - أ.(∇ x ب)

i.e: $\text{div}\ \bar{A}\times\vec{B} = \bar{B}\cdot\text{curl}\,\bar{A} - \bar{A}\cdot\text{curl}\,B$

$\nabla\cdot$ 5

$$\boxed{\nabla\cdot(\bar{A}\times\bar{B}) = \bar{B}\cdot(\nabla\times\bar{A}) - \bar{A}\cdot(\nabla\times\bar{B})}$$

L.H.S $\nabla\cdot(\bar{A}\times\bar{B}) = \left(\frac{\partial}{\partial x}i + \frac{\partial}{\partial y}j + \frac{\partial}{\partial z}k\right)\cdot(\bar{A}\times\bar{B})$

$$= \left(\frac{\partial}{\partial x}i + \frac{\partial}{\partial y}j + \frac{\partial}{\partial z}k\right)\cdot\begin{vmatrix} i & j & B \\ A_x & A_y & A_z \\ B_x & B_y & B_z \end{vmatrix}$$

$$= \begin{vmatrix} \frac{\partial}{\partial x} & \frac{\partial}{\partial y} & \frac{\partial}{\partial z} \\ A_x & A_y & A_z \\ B_x & B_y & B_z \end{vmatrix}$$

$$= \frac{\partial}{\partial x}(A_yB_z - B_yA_z) + \frac{\partial}{\partial y}(B_xA_z - A_xB_z) + \frac{\partial}{\partial z}(A_xB_y - B_xA_y)$$

$$= \left[\left(A_y\frac{\partial B_z}{\partial x} + B_z\frac{\partial A_y}{\partial x}\right) - \left(B_y\frac{\partial A_z}{\partial x} + A_z\frac{\partial B_y}{\partial x}\right)\right] +$$

$$\left[\left(B_x \frac{\partial A_z}{\partial y} + A_z \frac{\partial B_x}{\partial y}\right) - \left(A_x \frac{\partial B_z}{\partial y} + B_z \frac{\partial A_x}{\partial y}\right)\right] +$$

$$\left[\left(B_y \frac{\partial A_x}{\partial z} + A_x \frac{\partial B_y}{\partial z}\right) - \left(B_x \frac{\partial A_y}{\partial z} + A_y \frac{\partial B_x}{\partial z}\right)\right]$$

$$\bullet = B_x\left(\frac{\partial A_z}{\partial y} - \frac{\partial A_y}{\partial z}\right) + B_y\left(\frac{\partial A_x}{\partial z} - \frac{\partial A_z}{\partial x}\right) + B_z\left(\frac{\partial A_y}{\partial x} - \frac{\partial A_x}{\partial y}\right)$$

$$-\left[A_x\left(\frac{\partial B_z}{\partial y} - \frac{\partial B_y}{\partial z}\right) + A_y\left(\frac{\partial B_x}{\partial z} - \frac{\partial B_z}{\partial x}\right) + A_z\left(\frac{\partial B_y}{\partial x} - \frac{\partial B_x}{\partial y}\right)\right]$$

$$\bullet = \bar{B}\cdot(\nabla \times \bar{A}) - \bar{A}\cdot(\nabla \times \bar{B})$$

i.e: div curl $\bar{A} = 0$

VI
6

$$\boxed{\nabla\cdot(\nabla \times \bar{A}) = 0}$$

This means that the curl of a vector field is itself a solenoidal field because if $\nabla\cdot\bar{V} = 0$ $\bar{V}$ is called soleniodal or tube. جال ذا خط

$$\bullet\ \nabla\cdot(\nabla \times \bar{A}) = \begin{vmatrix} \frac{\partial}{\partial x} & \frac{\partial}{\partial y} & \frac{\partial}{\partial z} \\ \frac{\partial}{\partial x} & \frac{\partial}{\partial y} & \frac{\partial}{\partial z} \\ A_x & A_y & A_z \end{vmatrix} = \frac{\partial}{\partial x}\left(\frac{\partial A_z}{\partial y} - \frac{\partial A_y}{\partial z}\right) + \frac{\partial}{\partial y}\left(\frac{\partial A_x}{\partial z} - \frac{\partial A_z}{\partial x}\right) + \frac{\partial}{\partial z}\left(\frac{\partial A_y}{\partial x} - \frac{\partial A_x}{\partial y}\right)$$

$$= 0$$

Assuming contgneous second partial div each bracket equal zero.

VII
7

$$\boxed{\nabla \cdot (\nabla \phi_1 \times \nabla \phi_2) = 0}$$

i.e div(grad ϕ_1 x grad ϕ_2) = 0

L.H.S Scalar div product

$$= \begin{vmatrix} \frac{\partial}{\partial x} & \frac{\partial}{\partial y} & \frac{\partial}{\partial z} \\ \frac{\partial \phi_1}{\partial x} & \frac{\partial \phi_1}{\partial y} & \frac{\partial \phi_1}{\partial z} \\ \frac{\partial \phi_2}{\partial x} & \frac{\partial \phi_2}{\partial y} & \frac{\partial \phi_2}{\partial z} \end{vmatrix} = 0$$

as can be easily proved by differentiation.

Note:

$$\begin{vmatrix} \frac{\partial}{\partial x} & \frac{\partial}{\partial y} & \frac{\partial}{\partial y} \\ \frac{\partial \phi_1}{\partial x} & \frac{\partial \phi_2}{\partial y} & \frac{\partial \phi_1}{\partial z} \\ \frac{\partial \phi_2}{\partial x} & \frac{\partial \phi_2}{\partial y} & \frac{\partial \phi_2}{\partial z} \end{vmatrix}$$

$$= \frac{\partial}{\partial x}\left(\frac{\partial \phi_1}{\partial y} \cdot \frac{\partial \phi_2}{\partial z} - \frac{\partial \phi_1}{\partial z} \frac{\partial \phi_2}{\partial y}\right) +$$

$$\frac{\partial}{\partial y}\left(\frac{\partial \phi_2}{\partial x} \cdot \frac{\partial \phi_1}{\partial z} - \frac{\partial \phi_1}{\partial x} \cdot \frac{\partial \phi_2}{\partial z}\right) +$$

$$\frac{\partial}{\partial z}\left(\frac{\partial \phi_1}{\partial x} \cdot \frac{\partial \phi_2}{\partial y} - \frac{\partial \phi_2}{\partial x} \cdot \frac{\partial \phi_1}{\partial y}\right)$$

$$= \left(\frac{\partial \Phi_1}{\partial y}\frac{\partial^2 \Phi_2}{\partial x \partial z} + \frac{\partial \Phi_2}{\partial z}\frac{\partial^2 \Phi_1}{\partial x \partial y}\right) - \left(\frac{\partial \Phi_1}{\partial z}\frac{\partial^2 \Phi_2}{\partial y \partial x} + \frac{\partial \Phi_2}{\partial y}\frac{\partial^2 \Phi_1}{\partial x \partial z}\right)$$

$$+ \left(\frac{\partial \Phi_2}{\partial x}\frac{\partial^2 \Phi_1}{\partial z \partial y} + \frac{\partial \Phi_1}{\partial z}\frac{\partial^2 \Phi_2}{\partial x \partial y}\right) - \left(\frac{\partial \Phi_1}{\partial x}\frac{\partial^2 \Phi_2}{\partial z \partial y} + \frac{\partial \Phi_2}{\partial z}\frac{\partial^2 \Phi_1}{\partial x \partial y}\right)$$

$$+ \left(\frac{\partial \Phi_1}{\partial x}\frac{\partial^2 \Phi_2}{\partial z \partial y} + \frac{\partial \Phi_2}{\partial y}\frac{\partial^2 \Phi_1}{\partial x \partial z}\right) - \left(\frac{\partial \Phi_2}{\partial x}\frac{\partial^2 \Phi_1}{\partial z \partial y} + \frac{\partial \Phi_1}{\partial y}\frac{\partial^2 \Phi_2}{\partial x \partial z}\right)$$

$$= 0$$

VIII. $\nabla \times (\Phi \bar{A}) = \Phi(\nabla \times \bar{A}) + (\nabla \Phi) \times \bar{A}$

8

i.e.: $\text{curl}\, \Phi\bar{A} = \Phi\, \text{curl}\bar{A} + \text{grad}\, \Phi \times \bar{A}$

$$\nabla \times (\Phi \bar{A}) = \begin{vmatrix} i & j & k \\ \frac{\partial}{\partial x} & \frac{\partial}{\partial y} & \frac{\partial}{\partial z} \\ \Phi A_x & \Phi A_y & \Phi A_z \end{vmatrix}$$

$$= \left[\frac{\partial}{\partial y}(\Phi A_z) - \frac{\partial}{\partial z}(\Phi A_y)\right] i$$

$$+ \left[\frac{\partial}{\partial z}(\Phi A_x) - \frac{\partial}{\partial x}(\Phi A_z)\right] j$$

$$+ \left[\frac{\partial}{\partial x}(\Phi A_y) - \frac{\partial}{\partial y}(\Phi A_x)\right] k$$

$$\bullet = \left[\left(\phi\frac{\partial A_z}{\partial y} + A_z\frac{\partial \phi}{\partial y}\right) - \left(\phi\frac{\partial A_y}{\partial z} + A_y\frac{\partial \phi}{\partial z}\right)\right] i$$

$$+ \left[\left(\phi\frac{\partial A_x}{\partial z} + A_x\frac{\partial \phi}{\partial z}\right) - \left(\phi\frac{\partial A_z}{\partial x} + A_z\frac{\partial \phi}{\partial x}\right)\right] j$$

$$+ \left[\left(\phi\frac{\partial A_y}{\partial x} + A_y\frac{\partial \phi}{\partial x}\right) - \left(\phi\frac{\partial A_x}{\partial y} + A_x\frac{\partial \phi}{\partial y}\right)\right] k$$

$$\bullet = A_x\left(\frac{\partial \phi}{\partial z} j - \frac{\partial \phi}{\partial y} k\right) + A_y\left(\frac{\partial \phi}{\partial x} k - \frac{\partial \phi}{\partial z} i\right) + A_z\left(\frac{\partial \phi}{\partial y} i - \frac{\partial \phi}{\partial x} j\right)$$

$$+ \phi\left[\left(\frac{\partial A_z}{\partial y} - \frac{\partial A_y}{\partial z}\right) i + \left(\frac{\partial A_x}{\partial z} - \frac{\partial A_z}{\partial x}\right) j + \left(\frac{\partial A_y}{\partial x} - \frac{\partial A_x}{\partial y}\right) k\right]$$

$$= \left(A_z\frac{\partial \phi}{\partial y} - A_y\frac{\partial \phi}{\partial z}\right) i + \left(A_x\frac{\partial \phi}{\partial z} - A_z\frac{\partial \phi}{\partial x}\right) j + \left(A_y\frac{\partial \phi}{\partial x} - A_x\frac{\partial \phi}{\partial y}\right) k$$

$$+ \phi\left[\left(\frac{\partial A_z}{\partial y} - \frac{\partial A_y}{\partial z}\right) i + \left(\frac{\partial A_x}{\partial z} - \frac{\partial A_z}{\partial x}\right) j + \left(\frac{\partial A_y}{\partial x} - \frac{\partial A_x}{\partial y}\right) k\right]$$

$$= \begin{vmatrix} i & j & k \\ -A_x & -A_y & -A_z \\ \frac{\partial \phi}{\partial x} & \frac{\partial \phi}{\partial y} & \frac{\partial \phi}{\partial z} \end{vmatrix} + \phi \begin{vmatrix} i & j & k \\ -A_x & -A_y & -A_z \\ \frac{\partial}{\partial x} & \frac{\partial}{\partial y} & \frac{\partial}{\partial z} \end{vmatrix}$$

From the rule of determinant, change the rows and change the sign of the determinant.

$$\therefore \nabla\times(\phi\bar{A}) = \begin{vmatrix} i & j & k \\ \frac{\partial\phi}{\partial x} & \frac{\partial\phi}{\partial y} & \frac{\partial\phi}{\partial z} \\ A_x & A_y & A_z \end{vmatrix} + \phi \begin{vmatrix} i & j & k \\ \frac{\partial}{\partial x} & \frac{\partial}{\partial y} & \frac{\partial}{\partial z} \\ A_x & A_y & A_z \end{vmatrix}$$

$$= \nabla\phi\times\bar{A} + \phi(\nabla\times\bar{A})$$

IX. $\boxed{\nabla\times(\nabla\phi) = 0}$

9

i.e: curl grad $\phi = 0$

That is the gradient of a scalar field is an irrotational Field.

$$\nabla\times(\nabla\phi) = \begin{vmatrix} i & j & k \\ \frac{\partial}{\partial x} & \frac{\partial}{\partial y} & \frac{\partial}{\partial z} \\ \frac{\partial\phi}{\partial x} & \frac{\partial\phi}{\partial y} & \frac{\partial\phi}{\partial z} \end{vmatrix} = \left(\frac{\partial^2\phi}{\partial z\partial y} - \frac{\partial^2\phi}{\partial y\partial z}\right)i + \left(\frac{\partial^2\phi}{\partial x\partial z} - \frac{\partial^2\phi}{\partial z\partial x}\right)j + \left(\frac{\partial^2\phi}{\partial y\partial x} - \frac{\partial^2\phi}{\partial x\partial y}\right)k$$

Assuming contaneous second partial drivative

X. 10

$$\boxed{\nabla \cdot \nabla \phi = \nabla^2 \phi}$$

i.e: div grad $\phi = \nabla^2 \phi$

$$\begin{aligned} L.H.S &= \nabla \cdot (\nabla \phi) \\ &= \frac{\partial}{\partial x}\left(\frac{\partial \phi}{\partial x}\right) + \frac{\partial}{\partial y}\left(\frac{\partial \phi}{\partial y}\right) + \frac{\partial}{\partial z}\left(\frac{\partial \phi}{\partial z}\right) \\ &= \frac{\partial^2 \phi}{\partial x^2} + \frac{\partial^2 \phi}{\partial y^2} + \frac{\partial^2 \phi}{\partial z^2} \\ &= \nabla^2 \phi \end{aligned}$$

XI 11

$$\boxed{\nabla \times (\nabla \times \bar{A}) = \nabla(\nabla \cdot \bar{A}) - \nabla^2 \bar{A}}$$

i.e curl curl $\bar{A}$ = grad div $\bar{A} - \nabla^2 \bar{A}$

$\because \bar{a} \times (\bar{b} \times \bar{c}) = \bar{b}\,(\bar{a} \cdot \bar{c}) - \bar{c}\,(\bar{a} \cdot \bar{b})$

scalar   scalar

or

$\bar{a} \times (\bar{b} \times \bar{c}) = (\bar{a} \cdot \bar{c})\bar{b} - (\bar{a} \cdot \bar{b})\bar{c}$

1st 2nd 3rd   1st 3rd 2nd – 1st 2nd 3rd

132   123

$$\therefore \nabla \times (\nabla \times \bar{A}) = (\nabla \cdot \bar{A})\nabla - (\nabla \cdot \nabla)\bar{A}$$

$$= \nabla(\nabla \cdot \bar{A}) - \nabla^2 \bar{A}$$

16. Examples

16.1. Directional derivatives

I - Calulate the directional derivative of $\emptyset = xy^2 + yz^3$ at at (2, - I ,I) in the direction of $2\underline{i} + \underline{j} + 2\underline{k}$. What is the maximum directional derivative at that point?

$$\nabla \emptyset = \nabla (xy^2 + yz^3) = y^2 \underline{i} + (2xy + z^3)\underline{j} + 3yz^2 \underline{k}$$

$$= i - 3j - 3k \qquad \text{at } (2,-1,-1)$$

The unit vector in the direction of $2\underline{i} + \underline{j} + 2\underline{k}$ is

$$\underline{a} = \frac{2}{3}\underline{i} + \frac{1}{3}\underline{j} + \frac{2}{3}\underline{k}$$

Then the required directional derivative is

$$\nabla \emptyset . \underline{a} = (\underline{I} - 3\underline{j} - 3\underline{k}) . (-\frac{2}{3}\underline{i} + \frac{I}{3}\underline{j} + -\frac{2}{3}\underline{k}) = \frac{2}{3} - 1 - 2 = -\frac{7}{3} .$$

The directional derivative is maximum in the direction of

$$\nabla \emptyset = \underline{i} - 3\,\underline{i} - 3\underline{k} .$$

The magnitude of the maximum is $|\nabla \emptyset| = \sqrt{1 + 9 + 9} = \sqrt{19}$.

16.2. Surface equation

2- Find the acute angle between the surfaces

$xy^2z = 3x+z^2$ and $3x^2-y^2+2z=1$ at the point (1,-2,1).

The angle between the surfaces at the point is the angle between the normals to the surfaces at the point. The normal to $3x - xy^2z + z^2 = 0$ at (1,-2,1) is along

$$\nabla \phi_1 = \nabla(3x - xy^2z + z^2) = -\underline{i} + 4\underline{j} - 2\underline{k}.$$

The normal to $3x^2 - y^2 - 2z = 1$ at (1,-2,1) is along

$$\nabla \phi_2 = \nabla(3x^2 - y^2 + 2z) = 6\underline{i} + 4\underline{j} + 2\underline{k}$$

$$(\nabla\phi_1)\cdot(\nabla\phi_2) = |\nabla\phi_1|\,|\nabla\phi_2|\cos\theta$$

$$\cos\theta = \frac{(-\underline{i}+4\underline{j}-2\underline{k})\cdot(6\underline{i}+4\underline{j}+2\underline{k})}{\sqrt{1+16+4}\,\sqrt{36+16+4}} = \frac{-6+16-4}{\sqrt{21}\sqrt{56}} = \frac{5}{7\sqrt{6}}$$

16.3. Solendoidal vectors

3 - Prove that the vector

$\underline{A} = 3y^4z^2\,\underline{i} + 4x^3z^2\,\underline{j} - 3x^2y^2\,\underline{k}$ is solenodical.

A vector is solenoidal if its divergence is zero

Then $$\nabla\cdot\underline{A} = \frac{\partial}{\partial x}(3y^4z^2) + \frac{\partial}{\partial y}(4x^3z^2) - \frac{\partial}{\partial z}(3x^2y^2) = 0.$$

16.4. Conservative force field

4 - Prove that $\underline{F} = r^2\underline{r}$ is a conservative force field. Find the scalar potential for $\underline{F}$, and the work done in moving an object in this field from (0,2,-1) to (1,2,-2)

$$\underline{F} = r^2\underline{r} = (x^2 + y^2 + z^2)(\underline{i}x + \underline{j}y + \underline{k}z).$$

A necessary and sufficient condition that $\underline{F}$ will be conservative is that curl $\underline{F} = \nabla \wedge \underline{F} = 0$.

$$\text{Now } \nabla \wedge \underline{F} = \begin{vmatrix} \underline{i} & \underline{j} & \underline{k} \\ \frac{\partial}{\partial x} & \frac{\partial}{\partial y} & \frac{\partial}{\partial z} \\ x(x^2+y^2+z^2) & y(x^2+y^2+z^2) & z(x^2+y^2+z^2) \end{vmatrix} = 0$$

Thus $\underline{F}$ is a conservative force field.

$$\underline{F}.\, d\underline{r} = \nabla \phi .\, d\underline{r} = \frac{\partial \phi}{\partial x} dx + \frac{\partial \phi}{\partial y} dy + \frac{\partial \phi}{\partial z} dz = d\phi.$$

Then $d\phi = \underline{F}.d\underline{r} = (x^2 + y^2 + z^2)(xdx + ydy + zdz) = \frac{1}{4} d\left[(x^2+y^2+z^2)^2\right]$

$$\therefore \phi = \frac{(x^2+y^2+z^2)^2}{4} + \text{constant} = \frac{r^4}{4} + \text{cons.}$$

<u>Another method</u> $\underline{F} = \nabla \phi$ or $\frac{\partial \phi}{\partial x}\underline{i} + \frac{\partial \phi}{\partial y}\underline{j} + \frac{\partial \phi}{\partial z}\underline{k}$

$$= (x^2+y^2+z^2)(\underline{i}x+\underline{j}y+\underline{k}z)$$

$$\frac{\partial \phi}{\partial x} = x(x^2+y^2+z^2) \quad (I)$$

$$\frac{\partial \phi}{\partial y} = y(x^2+y^2+z^2) \quad (2)$$

$$\frac{\partial \phi}{\partial z} = z(x^2+y^2+z^2) \quad (3)$$

Integrating (I), (2) and (3), we get

$$\phi = \frac{x^4}{4} + \frac{x^2y^2}{2} + \frac{x^2z^2}{2} + f(y,z)$$

$$\phi = \frac{y^2x^2}{2} + \frac{y^4}{4} + \frac{y^2z^2}{2} + g(x,z)$$

$$\phi = \frac{x^2z^2}{2} + \frac{y^2z^2}{2} + \frac{z^4}{4} + h(x,y)$$

$$\therefore \phi = \frac{1}{4}\left[x^4+y^4+z^4+2x^2y^2+2y^2z^2+2z^2x^2\right] + \text{const}$$

$$= \tfrac{1}{4}(x^2+y^2+z^2)^2 + \text{const} = \frac{r^4}{4} + \text{const.}$$

$$\text{Work done} = \int_{P_I}^{P_2} \underline{F} \cdot d\underline{r} = \int_{P_I}^{P_2} d\phi = -\frac{I}{4} \int_{(0,2,-I)}^{(I,2,-2)} d(x^2+y^2+z^2)^2$$

$$= \phi\,(I,2,-2\,) - \phi\,(o,2,-I) = \frac{I}{4}\,\left[8I - 25\right] = 14.$$

16.5. Linear integral of a vector function

5. A vector function $\underline{V}$ is given by:

$$\underline{V} = X(x)\, Y(y) \left\{ \underline{i} \cos z + \underline{j}\, Z(z) + \underline{k} \sin z \right\}$$

where $X(0) = Y(0) = Z(0) = 1$. If for any two points A and B,

$\int_A^B \underline{V}.d\underline{r}$ is independent of the path of integration, find X, Y, Z.

Hence, evaluate the integral when A is at the origin and B is the point $(1, 1, \frac{\pi}{2})$.

Since $\int_A^B \underline{V}.d\underline{r}$ is independent of the path of integration, therefore the vector function $\underline{V}$ is conservative

$$\therefore \quad \nabla \wedge V = \begin{vmatrix} \underline{i} & \underline{j} & \underline{k} \\ \frac{\partial}{\partial x} & \frac{\partial}{\partial y} & \frac{\partial}{\partial z} \\ XY \cos z & XYZ & XY \sin z \end{vmatrix}$$

$$\therefore \quad \sin Z \frac{dY}{dy} - Y \frac{dZ}{dz} = 0 \; ; \qquad (1)$$

$$\frac{dX}{dx} + X = 0 \; ; \qquad (2)$$

$$YZ \frac{dX}{dx} - X \cos z \frac{dY}{dy} = 0 \, . \qquad (3)$$

Integrating equation (2) and since $X(0) = 1$, we get:

$$X = e^{-x} .$$

Equation (3) becomes

$$\cos z \frac{dY}{dy} + YZ = 0 \quad . \tag{4}$$

From (1) and (4)

$$\frac{dZ}{dz} + \tan z \;\; Z = 0 .$$

This gives on integration

$$Z = A \cos z.$$

But $Z(0) = -1$, therefore $A = -1$ and

$$Z = - \cos z$$

Substitute in (4) we obtain:

$$\frac{dY}{dy} - Y = 0$$

i.e. $\qquad Y = e^y ,$

since $Y(0) = 1$.

$$\int_{0,0,0}^{1,1,\frac{\pi}{2}} \underline{V}.d\underline{r} = \int_{0,0,0}^{1,1,\frac{\pi}{2}} e^{y-x}(\cos z\, dx - \cos z\, dy + \sin z\, dz)$$

$$= \int_{0,0,0}^{1,1,\frac{\pi}{2}} e^{y-x}(-\cos z\, d(y-x) + \sin z\, dz)$$

$$= - \int_{0,0,0}^{1,1,\frac{\pi}{2}} d(\cos z\, e^{y-x})$$

$$= - \left[\cos z\, e^{y-x}\right]_{0,0,0}^{1,1,\frac{\pi}{2}}$$

$$= 1 .$$

16.6. Solenoidal vector

6. Determine the constant a so that the vector

$$\underline{V} = (x + 3y)\, \underline{i} + (y - 2z)\, \underline{j} + (x + az)\, \underline{k}$$

is solenoidal. Calculate a vector potential of $\underline{V}$.

In order that $\underline{V}$ be solenoidal, we have

$$\text{div } \underline{V} = 1 + 1 + a = 0$$

$$\text{i.e.} \quad a = -2.$$

Assume that

$$\underline{V} = \text{curl } \underline{F} \tag{1}$$

where

$$F_x = \int V_y \, dz = z(y - z) + f(x,y) ,$$

$$F_y = - \int V_x \, dz = -z(x + 3y) + f'(x,y) ,$$

$$F_z = 0 .$$

Substitute in (1)

$$\underline{V} = \begin{vmatrix} \underline{i} & \underline{j} & \underline{k} \\ \frac{\partial}{\partial x} & \frac{\partial}{\partial y} & \frac{\partial}{\partial z} \\ z(y-z) + f & -z(x+3y) + f' & 0 \end{vmatrix}$$

$$\therefore \quad \frac{\partial f'}{\partial x} - \frac{\partial f}{\partial y} = x ,$$

This can be satisfied by choosing $f = 0$

and $f' = \frac{1}{2} x^2$

Also an arbitrary function in the form $\nabla \phi$ can be added to $\underline{F}$ since $\text{curl} \nabla \phi = 0$.

$\therefore$ A vector potential of $\underline{V}$ can be taken to be:

$$\underline{F} = z(y - z)\underline{i} + (\tfrac{1}{2} x^2 - zx - 3zy)\underline{j} + \nabla \phi .$$

17. Exercises on vector analysis

1) Find the angle between the two surfaces

$$x^2 + y^2 + z^2 = 9 \quad \text{and} \quad z = x^2 + y^2 - 3$$

at the point (2, -1, 2).

2) In what direction from the point (2, 1, -1) is the directional derivative of $\phi = x^2 y z^3$ is maximum?

Find the magnitude of this maximum.

3) Find the directional derivative of $\phi = x^2 y z + 4xz^2$ at (1, -2, -1) in the direction of $2\underline{i} - \underline{j} - 2\underline{k}$. What is the maximum directional derivative at that point?

4) (a) Calculate the magnitude of the maximum rate of change of $\phi = z^2(x^2 - y^2)$ at the point (1, 2, 3).

(b) Find the co-ordinates of the points on the plane $z = 1$ at which the direction of the maximum rate of change of ϕ in (a) is parallel to the vector $\underline{i} - 2\underline{j} - 3\underline{k}$.

(c) Write down the equation of the level surface for ϕ in (a) that passes through the point (cosh t, sinh t, c) where t and c are constant.

5) Calculate the rate of change of $\phi = xy^2z^3$ in the direction of the normal to the surface $x^2 + xy^2 + yz^2 = 7$, at the point P(1, 2, 1).
Calculate the change in ϕ, in this direction, when the x-coordinate, measured from P changes by a very small amount δx.

6. Show that $\underline{F} = (2xy + z^3)\underline{i} + x^2\,\underline{j} + 3xz^2\underline{k}$ is a conservative force field and find its scalar potential. Calculate the work done in moving a particle in this field from (1,-2,1) to (3,1,4).

7. If $\underline{A} = (4xy - 3x^2z^2)\underline{i} + 2x^2\underline{j} - 2x^3z\underline{k}$, prove that $\int_C \underline{A}.d\underline{r}$ is independent of the curve C. Show that there is a differentiable function ϕ such that $\underline{A} = \nabla\phi$ and find it.

8. Prove that $\underline{F} = (y^2 \cos x + z^3)\underline{i} + (2y \sin x - 4)\underline{j} + (3xz^2 + 2)\underline{k}$ is a conservative force field. Find the potential ϕ and calculate the work done in moving a particle in this field form (0, 1, -1) to $(\frac{\pi}{2}, -1, 2)$.

9. Find a, b, c such that
$\underline{V} = (x + 2y + az)\underline{i} + (bx - 3y - z)\underline{j} + (4x + cy + 2z)\underline{k}$ is irrotational and find its scalar potential. Is $\underline{V}$ solenoidal?

10. Show that $\underline{q} = yz\,\underline{i} + xz\,\underline{j} + xy\,\underline{k}$ is solenoidal. Calculate a vector potential of $\underline{q}$. Prove that for any closed curve C, $\oint_C \underline{q}.\,d\underline{r} = 0.$

11. Show that if $\phi(x,y,z)$ is a solution of Leplace's equation, then $\nabla\phi$ is a vector which is both solenoidal and irrotational.

12. Considering $\nabla^2\phi$ as div (grad ϕ) show that

(a) $\nabla^2 f(r) = \frac{d^2f}{dr^2} + \frac{2}{r}\frac{df}{dr}$

(b) $\nabla^2(\phi\psi) = \phi\nabla^2\psi + 2\nabla\phi.\nabla\psi + \psi\nabla^2\phi.$

13. Prove that $\underline{F} = \underline{r}f(r)$, where $f(r)$ is a single valued function, is conservative. Find the potential when:

(a) $f(r) = r$ (b) $f(r) = \frac{1}{r^2}$

In each case, calculate $\int_{1,1,1}^{x,y,z} \underline{F}.\,d\underline{r}.$

Also find $f(r)$ if the field is solenoidal.

14. If $\underline{F} = \omega(r)\,\underline{k}\wedge\underline{r}$, $\underline{r} = \underline{i}\,x + \underline{j}y$, find $\omega(r)$ if the field is

(a) irrotational, (b) solenoidal.

15. If $\underline{A}$ and $\underline{B}$ are irrotational, prove that $\underline{A}\wedge\underline{B}$ is solenoidal.

16. A vector function $\underline{V}$ has components

$-\,xyz, \frac{1}{6}(x^2 - y^2)z \quad , \frac{1}{2}(x^2 + y^2)y.$

A scalar function $\phi(r)$ where $\phi(1) = 1$ and $r = (x^2 + y^2)^{1/2}$, is such that, for any two points A and B, $\int_A^B \phi\underline{V}.d\underline{r}$ is independent of the path of integration. Show that $\phi(r) = r^{-4}$ and evaluate the integral when A is (a,a,a) and B is (b,b,b).

17. A vector function $\underline{V}$ has components:

$e^z \cos y, \ e^z \sin y, \ e^z f.$

where f is a function of y only and f(0) = -1. A scalar function ϕ is a function of x only and $\phi(0) = 1$. If for any two points A and B, $\int_A^B \phi \underline{V}.d\underline{r}$ is independent of the path of integration, find f and ϕ. Hence, evaluate the integral when A is at the origin and B is at (a,b,c).

18. Determine which, if any, of the following vector fields are expressible in the form $\nabla \phi$:

(a) $(\underline{a}.\underline{r})\underline{b}$ (b) $(\underline{a}.\underline{b})\underline{r}$

(c) $(\underline{a}.\underline{r})\underline{b} + (\underline{b}.\underline{r})\underline{a}$ (d) $(\underline{a}.\underline{r})\underline{b} - (\underline{b}.\underline{r})\underline{a}$

where $\underline{a}$ and $\underline{b}$ are constant vectors. Find ϕ in the appropriate cases.

19. If $\underline{a}$ is a constant vector, prove that

$\nabla(\underline{a}.\underline{u}) = (\underline{a}.\nabla)\underline{u} + \underline{a} \wedge \text{curl } \underline{u},$

$\nabla\cdot(\underline{a} \wedge \underline{u}) = -\underline{a}.(\nabla \wedge \underline{u})$

$\nabla \wedge (\underline{a} \wedge \underline{u}) = \underline{a} \text{ div } \underline{u} - (\underline{a}.\nabla)\underline{u}.$

Hence, or otherwise, find the corresponding expressions when $\underline{u} = \underline{r}$.

20. If $\underline{a}$ is a constant vector, prove that:

$\nabla(\underline{a}.\underline{r})r^n = r^n \underline{a} + nr^{n-2}(\underline{a}.\underline{r})\underline{r}$

$\nabla.(\underline{a} \wedge \underline{r})r^n = 0$

$\nabla \wedge (\underline{a} \wedge \underline{r})r^n = (2+n)r^n \underline{a} - nr^{n-2}(\underline{a}.\underline{r})\underline{r}$

21. If $\underline{a}$, $\underline{b}$ are constant vectors, prove that:

(i) $\nabla(\underline{a}.\underline{r})(\underline{b}.\underline{r}) = (\underline{b}.\underline{r})\underline{a} + (\underline{a}.\underline{r})\underline{b}$,

(ii) $\underline{a}.\nabla(\underline{b}.\nabla\frac{1}{r}) = \frac{3(\underline{a}.\underline{r})(\underline{b}.\underline{r})}{r^5} - \frac{\underline{a}.\underline{b}}{r^3}$.

Also find the gradient of $\underline{a}.\nabla(\underline{b}.\nabla\frac{1}{r})$.

22. Given that $\rho\underline{F} = \nabla p$, where ρ, p, $\underline{F}$ are point functions,

prove that $\underline{F}.\ \text{curl}\ \underline{F} = 0$

If $p = f(\rho)$, then $\underline{F} = \nabla\int\frac{dp}{\rho}$.

23. P is any point on the ellipse whose focii are at the points A and B. Show that the lines PA and PB make equal angles with the tangent to the ellipse at P.

24. Express the inverse square law in terms of cylindrical coordinates (r, θ, z), the origin O being the centre of attraction. Is it conservative?

The point of application P of the force $-\frac{\overrightarrow{OP}}{OP^3} - \theta\,\hat{\underline{\theta}}$ moves rouhd the conical spiral given by the equation r = z, $\theta = z^2$ from z = 1 to z = 3. Calculate the work done by $\underline{F}$.

25. Two of Maxwells electromagnetic equations in rationalized M.K.S. units are:

div. $\underline{B} = 0$ and curl $\underline{E} = -\dfrac{\partial \underline{B}}{\partial t}$.

Given that the vector potential of $\underline{B}$ is $\underline{A}$, deduce that

$$\underline{E} + \frac{\partial \underline{A}}{\partial t} = -\text{grad } V,$$

where V is a scalar potential.

Given also that:

$$\text{curl } \underline{B} = \nu\left(\underline{i} + \alpha\frac{\partial \underline{E}}{\partial t}\right),$$

where ν is a constant, prove that:

$$\nabla^2\underline{A} - \nu\alpha\frac{\partial^2 \underline{A}}{\partial t^2} = -\nu\underline{i}$$

provided div $\underline{A}$ is chosen to equal $\left(-\nu\alpha\dfrac{\partial V}{\partial t}\right)$.

26. Given Maxwell's equations in empty space

$$\nabla \wedge \underline{H} = \frac{1}{c}\frac{\partial \underline{E}}{\partial t}, \quad \nabla \cdot \underline{H} = 0$$

$$\nabla \wedge \underline{E} = -\frac{1}{c}\frac{\partial \underline{H}}{\partial t}, \quad \nabla \cdot \underline{E} = 0,$$

where c is the constant velocity of light, show that $\underline{H}$ satisfies the equation

$$c^2 \nabla^2 \underline{H} = \frac{\partial^2 \underline{H}}{\partial t^2}.$$

If the vector field $\underline{Z}$ satisfies the equation

$$c^2 \nabla^2 \underline{Z} = \frac{\partial^2 \underline{Z}}{\partial t^2},$$

show that

$$\underline{E} = -\frac{1}{c}\frac{\partial}{\partial t}\nabla \wedge \underline{Z}, \quad \underline{H} = \nabla \wedge \nabla \wedge \underline{Z}$$

satisfy Maxwell's equations. If k and $\underline{a}$ are constant, evaluate the corresponding fields $\underline{E}$, $\underline{H}$, when

$$\underline{Z} = \frac{\underline{a}}{r} \cos k(r - ct).$$

27. The fundamental equation in elasticity theory for the displacement vector $\underline{D}$ when he body is at rest and the body force is zero, is

$$\text{grad div } \underline{D} + (1 - 2\eta) \nabla^2 \underline{D} = 0$$

where η is a constant. Show that this equation is satisfied by:

(i) $\underline{D} = \text{grad } \phi$ when $\nabla^2 \phi = 0$

(ii) $\underline{D} = z \text{ grad } \psi - (3 - 4\eta) \psi \underline{k}$, when $\nabla^2 \psi = 0$

CHAPTER 2: GAUSS'S DIVERGENCE THEOREM AND STOKES' THEOREM

1. Surface Integral

Let V be a scalar or vector point function which is defined and continuous on a surface S. Divide S into elements of area $\delta S_1, \delta S_2 \dots \delta S_i \dots \delta S_m$. Let V_i be the value of V at any point P_i within δS_i. Then the limit:

$$\lim_{m \to \infty} \sum_{i=1}^{m} V_i \, \delta S_i$$

which is usually abbreviated to

$$\lim_{\delta S \to 0} \Sigma V \delta S$$

is called the integral of V

over S and is written as $\int_S V \, dS$.

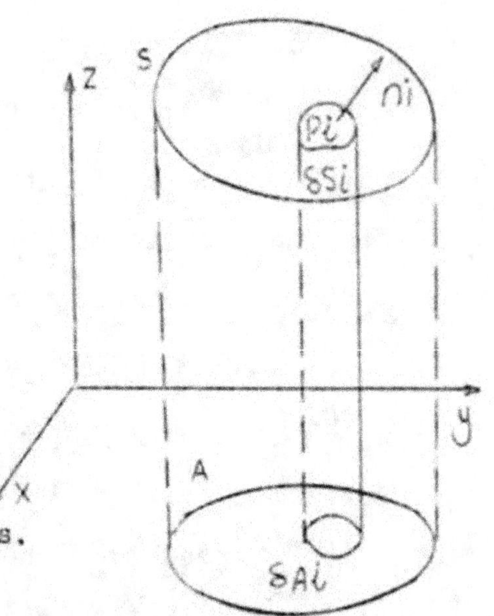

Surface integrals can always be transformed to others over plane areas in one of the coordinate planes. For let δA_i be the projection of

δS_i on the x, y-plane,

say;

i.e. $\delta A_i = |\underline{k}.\underline{n}_i| \; \delta S_i$

where $\underline{n}_i$ is the unit vector normal to S_i at P_i. Therefore

$$\int_S V \, dS = \lim_{m \to \infty} \sum_{i=1}^{\infty} V_i \, \delta S_i$$

$$= \lim_{m \to \infty} \sum_{i=1}^{\infty} \frac{V_i}{|\underline{k}.\underline{n}_i|} \delta A_i = \int_A \frac{V}{|\underline{k}.\underline{n}|} \, dA$$

where A is the projection of S on the x, y-plane.

This formula provides a convenient way to evaluate $\int_S VdS$, for, in most cases, it is simpler to evaluate $\int_A \frac{}{|\underline{k}.\underline{n}|} \, dA$. The normal to A must intersect S in one point only. This will produce no real difficulty as we can generally divide S into surfaces which do satisfy this restriction.

The particular surface integral $\int_S \underline{V}.\underline{n}\, dS = \int_S \underline{V}.d\underline{S}$ appears frequently in applied mathematics and physics. It is called the normal surface integral or flux of $\underline{V}$ over S.

1.1. Surface integral as scalar product

Surface integral as scalar Products

Let S be a surface drawn in a vector field F. P is any point on S. Surround P by an infinitesimal area ds. F is the vector through P. n is the unit vector at P normal to the surface outwards the surface integral of F over the surface S is defind as.

$$\iint_S \bar{F} \cdot n \, ds = \iint_S F \cos\theta \, ds$$

This means that we must multiply the scalar component of F normal to the surface by the area & then integrate.

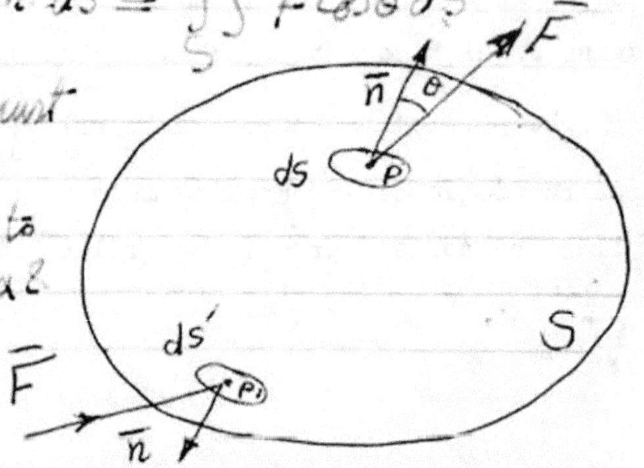

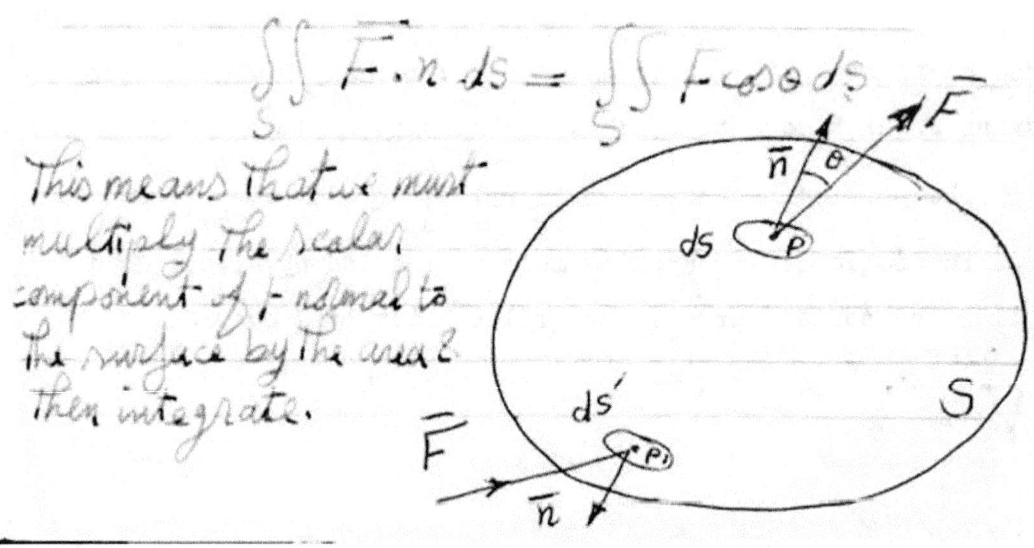

At the point like P_1, $\bar{F} \bar{n} \, ds$ is negative. if it happen that the surface integral is zero over a closed surface then this would means that the flux diverging is exactly balanced by equal flux converging on other point of the surface

$$\text{if} \quad \oiint_S \bar{F} \hat{n} \, ds = 0$$

$\therefore$ Flux diverging = Flux converging

If we defind the vector area $d\bar{s}$ as a vector normal to the area of magnitude equal to the area ds then the surface integral can be written in the form

$$\iint_S \bar{F} \cdot d\bar{s}$$

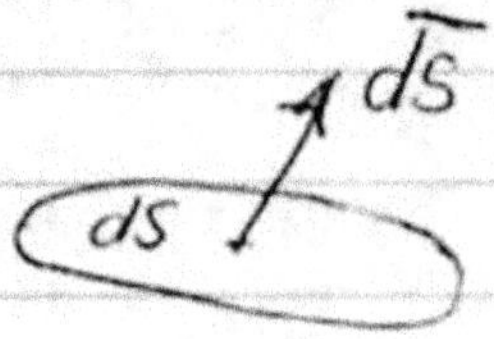

2. Volume Integrals

Let V be a scalar or vector point function defined and continuous in a volume τ.

Divide τ into elements of volume $\delta\tau_1, \delta\tau_2, \ldots \delta\tau_i, \ldots \delta\tau_n$.

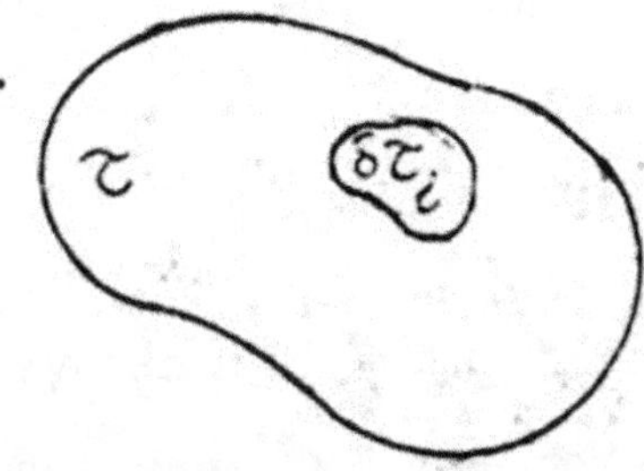

Let V_i be the value of V at any point within $\delta\tau_i$. Then the limit

$$\lim_{m\to\infty} \sum_{i=1}^{m} V_i \delta\tau_i$$

which is usually abbreviated to

$$\lim_{\delta\tau\to 0} \sum V \, \delta\tau$$

is called the volume integral of V throughout τ and is written as $\int_\tau V d\tau$.

3. Gauss's Divergence Theorem

If $\underline{V}$ is a vector point function which together with its derivative in any direction is single valued, finite and continuous in the region τ bounded by a closed surface S, then the outward flux of $\underline{V}$ over S is equal to the volume integral of the divergence of $\underline{V}$ taken throughout the enclosed volume.

If $\underline{n}$ is the unit vector along the outward normal, this theorem can be expressed in the form

$$\int_S \underline{V} \cdot \underline{n} \, dS = \int_\tau \operatorname{div} \underline{V} d\tau$$

To prove the theorem, divide τ into small elements by planes parallel to the coordinate planes distant δx, δy, δz apart respectively. Consider one of these elements; it is a rectangular parallelepiped of dimensions $\delta x, \delta y, \delta z$. Let its centre be at P(x,y,z).

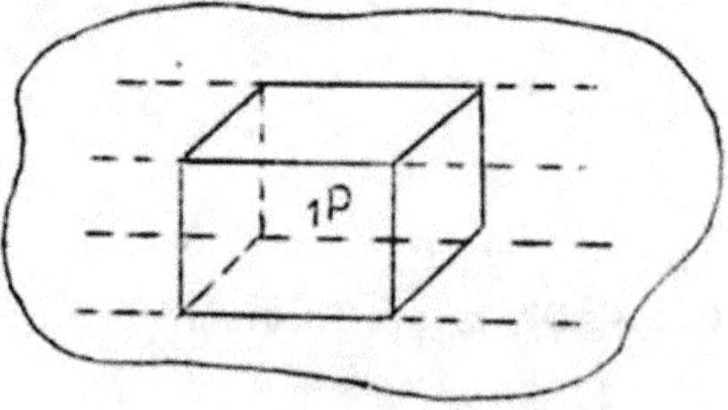

If $\underline{V}$ is the value of the vector function at P, its value at the centre of the sides of the element normal to the x-axis are:

$$\underline{V} - \frac{1}{2}\frac{\partial \underline{V}}{\partial x}\delta x, \ \underline{V} + \frac{1}{2}\frac{\partial \underline{V}}{\partial x}\delta x$$

The values can be taken to be the values of the vector function at each point of the corresponding side. Thefefore, the outward flux over these two sides:

$$= (V_x + \frac{1}{2}\frac{\partial V_x}{\partial x}\delta x)\delta y \delta z - (V_x - \frac{1}{2}\frac{\partial V_x}{\partial x}\delta x)\delta y \delta z$$
$$= \frac{\partial V_x}{\partial x}\delta \tau$$

Similarly, the outward flux over the sides normal to the y- and z-axes are

$$\frac{\partial V_y}{\partial y}\delta\tau, \quad \frac{\partial V_z}{\partial z}\delta\tau$$

respectively. Therefore, the outward flux over the element is

$$\int_{element} \underline{V}.\underline{n}dS = (\frac{\partial V_x}{\partial x} + \frac{\partial V_y}{\partial y} + \frac{\partial V_z}{\partial z})\partial\tau = \text{div}\underline{V}.\delta\tau$$

Adding for two neighbouring elements, the integral taken over the common surface will cancel, since their outward normals are in opposite directions while $\underline{V}$ is assumed continuous, finite and single valued.

Adding for all such elements, we will be left with the integration over the surface S, i.e. $\int_S \underline{V}.\underline{n}\, dS$.

But at the same time, by adding for these elements, we are integrating, div $\underline{V}$ throughout τ.

$$\therefore \int_S \underline{V}.\underline{n}\, dS = \int_\tau \text{div } \underline{V}\, d\tau$$

Notice that the above proof can be used to extend the validity of Gauss's theorem to a region τ bounded by an outer surface S and one or more inner surfaces S'; i.e.

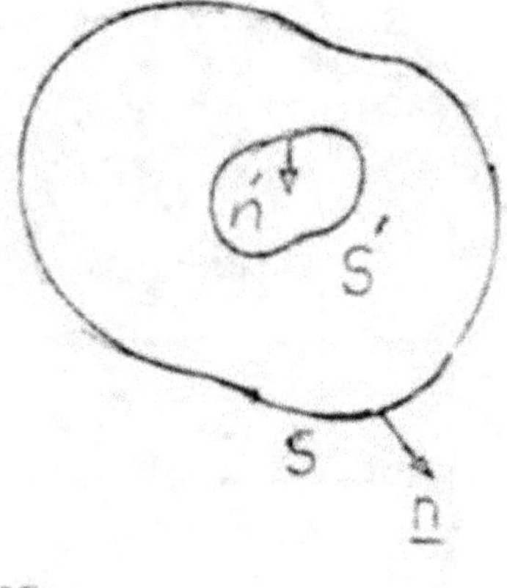

$$\int_\tau \text{div } \underline{V}\, d\tau = \int_S \underline{V}.\underline{n}dS + \int_{S'} \underline{V}.\underline{n}'dS$$

3.1. Relationships between divergence, volume, and surface of an enclosed space

The divergence theorem of gauss.

Let S be a closed surface including a volume V drawn in a vector field $\bar{F}$ then

$$\iiint_V (\nabla \cdot \bar{F})\, dv = \oiint_S \bar{F} \cdot \bar{n}\, ds$$

i.e: if we integrate the divergence over the volume V then this will be equal to

the surface integral of F over the surface S

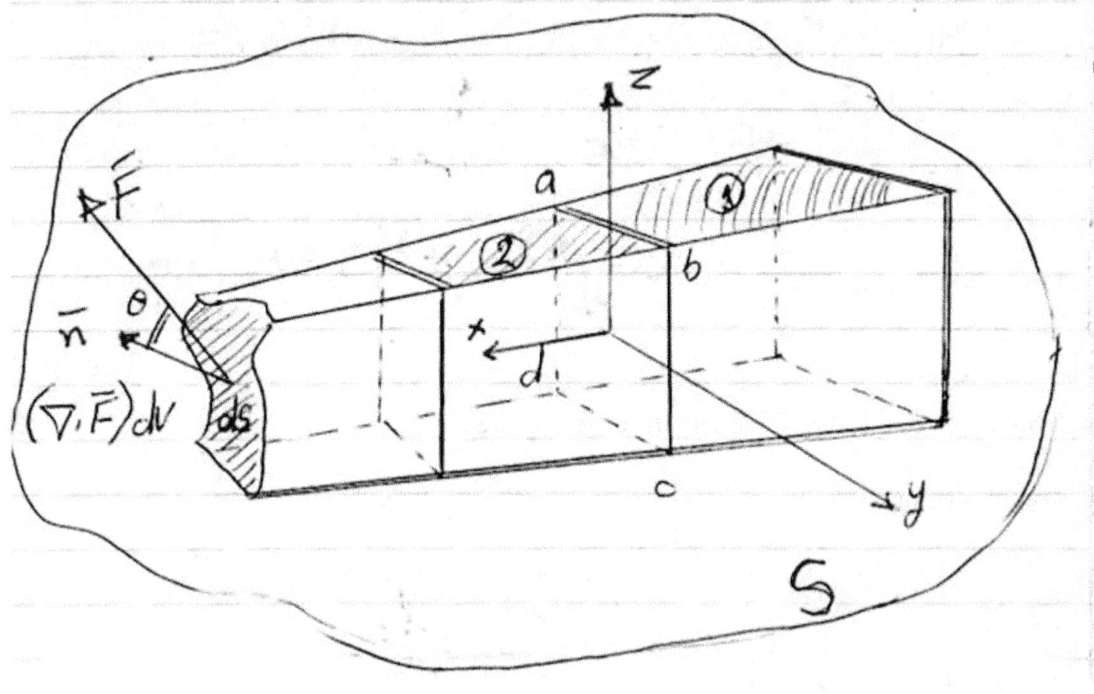

Subdivide the volume V into a large number of Rectangular parallelopipeds متوازيات سطوح in the x-y-z directions. Consider one adjacent elements ①, ② having a common face "abcd". Now we have already $(\nabla \cdot F)\,dv$ represents the flux diverging from the element of volume dV خ.ع

Now the flux diverging from element ① out of the surface ABCD is as the same time a flux converging on the element ② through the same face.

i.e. The flux inter twice with opposite signs. hence all such fluxes cancell one another the only fluxes which of unbalanced are those diverging from the surface whose aggregate are

$$\oiint_S \bar{F}\cdot\bar{n}\,ds$$

$$\therefore \iiint_V (\nabla\cdot\bar{F})\,dV = \oiint_S \bar{F}\cdot\bar{n}\,ds$$

4. Theorems deducible from the divergence theorem

<u>Constant Vector</u>

(i) Apply the divergence theorem to the function $\phi \underline{d}$, where ϕ is a scalar point function and $\underline{d}$ is a constant vector. Then:

$$\int_S \phi \underline{d}.\underline{n} dS = \int_\tau \text{div } \phi \underline{d}\, d\tau = \int_\tau \underline{d}.\nabla \phi d\tau$$

i.e. $\underline{d}. \int_S \phi \underline{n}\, dS = \underline{d}. \int_\tau \nabla \phi d\tau$

Since this is true for all $\underline{d}$, we must have:

$$\int_S \phi \underline{n} dS = \int_\tau \nabla \phi d\tau \qquad \text{(a)}$$

This vectorial equation is equivalent to the following three scalar equations:

$$\int_S \phi\, \underline{i}.\underline{n} dS = \int_\tau \frac{\partial \phi}{\partial x} d\tau ,$$

$$\int_S \phi\, \underline{j}.\underline{n} dS = \int_\tau \frac{\partial \phi}{\partial y} d\tau ,$$

$$\int_S \phi\, \underline{k}.\underline{n} dS = \int_\tau \frac{\partial \phi}{\partial z} d\tau$$

Variable Vector

(ii) Apply the divergence theorem to $\underline{V} \wedge \underline{d}$. Then as

$$\nabla \cdot (\underline{V} \wedge \underline{d}) = (\nabla \wedge \underline{V}) \cdot \underline{d} - (\nabla \wedge \underline{d}) \cdot \underline{V} = (\nabla \wedge \underline{V}) \cdot \underline{d} ,$$

$$n \cdot (\underline{V} \wedge d) = (\underline{n} \wedge \underline{V}) \cdot \underline{d},$$

The theorem leads to

$$\underline{d} \cdot \int_S \underline{n} \wedge V \, dS = \underline{d} \cdot \int_\tau \nabla \wedge \underline{V} \, d\tau$$

for all $\underline{d}$. We, then, must have

$$\int_S \underline{n} \wedge \underline{V} \, dS = \int_\tau \nabla \wedge \underline{V} \, d\tau \qquad \text{(b)}$$

(a) and (b) are two other forms of Gauss's theorem.

5. Infinitismal Volume with Closed Surface

If we apply the three forms of Gauss's theorem to an infinitely small closed surface S and the enclosed element of volume $\delta\tau$, we obtain:

$$\overline{\nabla \phi} = \frac{1}{\delta \tau} \int_S \underline{n} \, \phi \, dS,$$

$$\overline{\nabla \cdot \underline{V}} = \frac{1}{\delta \tau} \int_S \underline{n} \cdot V \, dS,$$

$$\overline{\nabla \wedge \underline{V}} = \frac{1}{\delta \tau} \int_S \underline{n} \wedge \underline{V} \, dS$$

where, for the function f, $\bar{f}$ denotes the average value of f in $\delta\tau$. Let the element shrink to a point, then $\bar{f}$ approaches the value of f at that point. Hence

$$\text{grad } \phi = \lim_{\delta\tau \to 0} \frac{1}{\delta\tau} \int_S \underline{n}\, \phi \, dS,$$

$$\text{div } \underline{V} = \lim_{\delta\tau \to 0} \frac{1}{\delta\tau} \int_S \underline{n}.\underline{V} \, dS,$$

$$\text{curl } \underline{V} = \lim_{\delta\tau \to 0} \frac{1}{\delta\tau} \int_S \underline{n} \wedge \underline{V} \, dS.$$

These lead to an alternative definition of grad ϕ, div $\underline{V}$ and curl $\underline{V}$ and since the values of the integrals in the above relations are independent of coordinate axes, it follows that these functions are invariant with respect to the choice of these.

6. Green's Theorem in the plane

Consider the two-dimensional field $\underline{V}$ whose components are P(x,y), -Q(x,y), 0. Apply Gauss's theorem to the right cylinder of unit height and whose base S lies in the x,y plane and is bounded by the curve C.

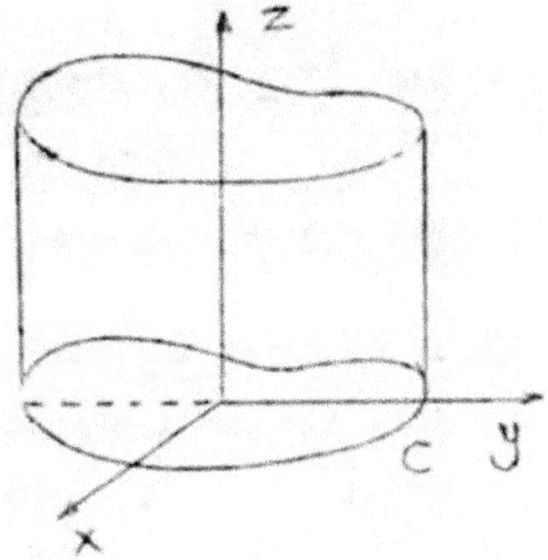

$$\therefore \int_{\tau} \text{div } \underline{V} \, d\tau = \int_{S} \left(\frac{\partial P}{\partial x} - \frac{\partial Q}{\partial y}\right) dS \times 1 .$$

$\int \underline{V}.\underline{n} \, dS = 0$ on the plane bases, since $\underline{V}$ has no component normal to them. For the curved side

$$\begin{aligned}\int \underline{V}.\underline{n} \, dS &= \oint_C \underline{V}.\underline{n} \, ds \times 1 \\ &= \oint_C (\underline{i}P - \underline{j}Q).(\underline{i}\frac{dy}{ds} - \underline{j}\frac{dx}{ds})ds \\ &= \oint_C P\,dy + Q\,dx .\end{aligned}$$

Therefore

$$\oint_C Q\,dx + P\,dy = \int_S \left(\frac{\partial P}{\partial x} - \frac{\partial Q}{\partial y}\right) dS .$$

This is known as Green's theorem in the plane. It also holds for multiply connected regions; i.e. regions bounded by an outer curve C and one or more inner curves C' which do not intersect. The above formula then takes the form

$$\oint_C Q\,dx + P\,dy + \oint_{C'} Q\,dx + P\,dy$$
$$= \int_S \left(\frac{\partial P}{\partial x} - \frac{\partial Q}{\partial y}\right) dx\,dy .$$

7. Green's Theorem

Let ϕ, ψ be two scalar point functions which together with their derivatives up to order 2 are single-valued, and continuous at all points of a region τ bounded by the closed surface S. Apply the divergence theorem for this region to the vector function $\phi \nabla \psi$. Therefore,

$$\int_\tau \operatorname{div} (\phi \nabla \psi) d\tau = \int_S \phi \nabla \psi . \underline{n} dS,$$

i.e.
$$\int_\tau (\nabla \phi . \nabla \psi + \phi \nabla^2 \psi) d\tau = \int_S \phi \frac{\partial \psi}{\partial n} dS,$$

which is Green's theorem.

Similarly we have

$$\int_\tau (\nabla \psi . \nabla \phi + \psi \nabla^2 \phi) d\tau = \int_S \psi \frac{\partial \phi}{\partial n} dS .$$

Subtracting the last two equations we get

$$\int_S (\phi \frac{\partial \psi}{\partial n} - \psi \frac{\partial \phi}{\partial n}) dS = \int_\tau (\phi \nabla^2 \psi - \psi \nabla^2 \phi) d\tau,$$

a symmetric form of Green's theorem.

8. Green's Formula

Let $\underline{r}$ be the position vector of any point relative to a point O. If ϕ and its derivatives up to order 2 at least are single valued and continuous in a region τ bounded by a surface S, then:

$$\int_S (\frac{1}{r} \frac{\partial \phi}{\partial n} - \phi \frac{\partial}{\partial n} \frac{1}{r}) dS - \int_\tau \frac{1}{r} \nabla^2 \phi d\tau = \alpha,$$

where $\alpha = 0$ or $4\pi\phi_0$ according as O is outside or inside S, ϕ_0 being the value of ϕ at O.

If O is out side S, then the function $\frac{1}{r}$ and its derivatives are single valued and continuous; also it is harmonic, i.e. $\nabla^2 \frac{1}{r} = 0$, at every point of τ. Applying the symmetric form of Green's theorem for the functions ϕ, $\frac{1}{r}$ to the region τ we get:

$$\int_S \left(\frac{1}{r}\frac{\partial \phi}{\partial n} - \phi \frac{\partial}{\partial n}\frac{1}{r}\right) dS - \int_\tau \frac{1}{r} \nabla^2 \phi \, d\tau = 0 .$$

If O is inside S, $\frac{1}{r}$ becomes infinite at O. We exclude this point from the region by a small sphere Σ of radius ε and centre O.

We can now apply Green's theorem to the region lying between S and Σ. Therefore,

$$\int_S \left(\frac{1}{r}\frac{\partial \phi}{\partial n} - \phi \frac{\partial}{\partial n}\frac{1}{r}\right) dS + \int_\Sigma \left(\frac{1}{r}\frac{\partial \phi}{\partial n} - \phi \frac{\partial}{\partial n}\frac{1}{r}\right) d\Sigma = \int \frac{1}{r} \nabla^2 \phi \, d\tau \quad \ldots. \quad (a)$$

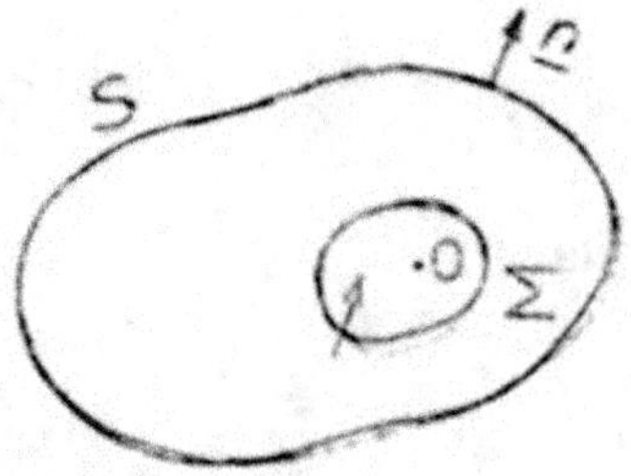

The surface integral over S is independent of Σ. For the integral over Σ we have:

$$\int_\Sigma \left(\frac{1}{r}\frac{\partial \phi}{\partial n} - \phi \frac{\partial}{\partial n}\frac{1}{r}\right) d\Sigma = -\int_\Sigma \left(\frac{1}{\epsilon}\frac{\partial \phi}{\partial r} + \frac{\phi}{\epsilon^2}\right) d\Sigma$$

$$= -\left(\frac{1}{\epsilon}\overline{\frac{\partial \phi}{\partial r}} + \frac{\bar{\phi}}{\epsilon^2}\right) 4\pi\epsilon^2$$

where as before $\bar{f}$ stands for the mean value of f over the sphere .

If $\epsilon \longrightarrow 0$, $\bar{f} \longrightarrow f_o$, and

$$\lim_{\epsilon \to 0} \int_\Sigma \left(\frac{1}{r}\frac{\partial \phi}{\partial n} - \phi \frac{\partial}{\partial n}\frac{1}{r}\right) d\Sigma = -4\pi \phi_o.$$

The volume integral applies to the region between S and Σ, but on proceeding to the limit, we take the whole region bounded by S. For, the magnitude of this difference is:

$$\left| \int_\Sigma \frac{1}{r} \nabla^2 \phi \, d\tau \right| \leq \int \frac{|\nabla^2 \phi|}{r} d\tau \leq M \int \frac{d\tau}{r}$$

$$= 2\pi M \epsilon^2 .$$

where M is the upper bound of $\nabla^2\phi$ inside Σ. The above magnitude will then vanish as $\epsilon \longrightarrow 0$.

Therefore equation (a) in the limit as $\epsilon \longrightarrow 0$ gives:

$$4\pi \phi_o = \int_S \left(\frac{1}{r}\frac{\partial \phi}{\partial n} - \phi \frac{\partial}{\partial n}\frac{1}{r}\right) dS - \int_\tau \frac{1}{r} \nabla^2 \phi \, d\tau .$$

Note:

If the point O lies on S, the surface Σ in the above proof is a hemisphere. In this case,

$$\lim_{\epsilon \to 0} \int_{\Sigma} \left(\frac{1}{r}\frac{\partial \phi}{\partial n} - \phi \frac{\partial}{\partial n}\frac{1}{r}\right) d\Sigma = -2\pi \phi_0,$$

thus

$$2\pi \phi_0 = \int_S \left(\frac{1}{r}\frac{\partial \phi}{\partial n} - \phi \frac{\partial}{\partial n}\frac{1}{r}\right) dS - \int_{\tau} \frac{1}{r} \nabla^2 \phi \, d\tau .$$

9. Green's formula for harmonic functions

When the function ϕ is harmonic, i.e. $\nabla^2 \phi = 0$, Green's formula becomes:

$$\int_S \left(\frac{1}{r}\frac{\partial \phi}{\partial n} - \phi \frac{\partial}{\partial n}\frac{1}{r}\right) dS = \alpha$$

where $\alpha = 0,\ 2\pi\phi_0,\ 4\pi\phi_0$ according to O is outside, on, or inside S.

10. Gauss's integral

Put ϕ equal to unity, then $\nabla^2 \phi = 0$, $\frac{\partial \phi}{\partial n} = 0$ and Green's formula leads to:

$$- \int_S \frac{\partial}{\partial n} \frac{1}{r} \, dS = - \int_S \underline{n} . \nabla \frac{1}{r} \, dS = \int_S \frac{\underline{n}.\underline{r}}{r^3} \, dS$$

$$\begin{aligned} &= 4\pi \quad \text{if O is inside S,} \\ &= 2\pi \quad \text{if O is on S,} \\ &= 0 \quad \text{if O is outside S.} \end{aligned}$$

The integral $\int_S \frac{\partial}{\partial n} \frac{1}{r} \, dS$ is known as Gauss's integral.

11. Solid angles

Let δS be an element of surface area in the direction $\underline{n}$ at the point whose vector position with respect to a point O is $\underline{r}$. Form the cone whose vertex is at O and whose base is δS. The projection of δS on a plane perpendicular to $\underline{r}$ is $\delta S \cos \theta$ where θ is the angle between δS and this plane. The area intercepted by the cone on the unit sphere with centre at O is

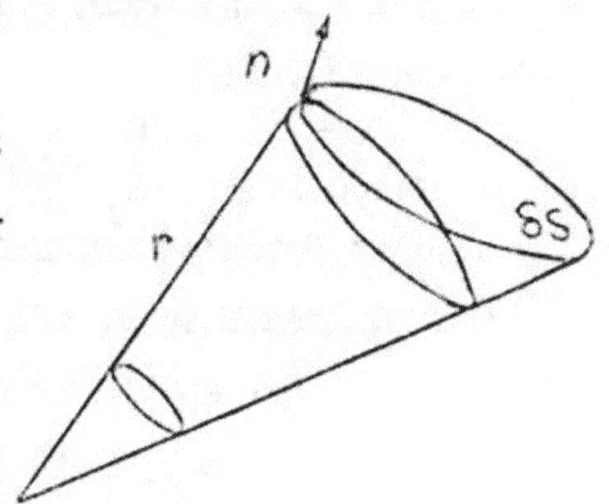

$$\frac{\cos \theta \, \delta S}{r^2}$$

The solid angle subtended by δS at O is defined to be this intercepted area and is positive or negative according, as the angle between $\underline{n}$ and $\underline{r}$ is acute or obtuse. It is usually denoted by δw, therefore.

$$\delta \omega = \frac{\underline{n}.\underline{r}}{r^3} \delta S$$

<u>Definition</u>:

The solid angle subtended at a point O by a surface S of any shape is measured by the area on the unit sphere with centre at O intercepted by the cone whose vertex is at O and whose base is S. i.e. for the surface S,

$$\omega = \int_S \frac{\underline{n}.\underline{r}}{r^3} dS .$$

For example, the solid angle subtended at the vertex of a right circular cone of vertical angle 2α has the value $2\pi(1-\cos\alpha)$, The solid angle subtended by an infinite plane at any point outside it has the value 2π.

12. Relation of Gauss's integral and the solid angle

For a closed surface S, the solid angle it subtends at a point O is

$$\int_S \frac{\underline{r}\cdot\underline{n}}{r^3}\, dS$$

which is Gauss's integral. If $\underline{n}$ is taken to be along the outward normal to S, then:

$$\begin{aligned} \omega &= 4\pi \text{ if O is inside S,} \\ &= 2\pi \text{ if O is on S,} \\ &= 0 \quad \text{if O is outside S.} \end{aligned}$$

i.e. if O crosses S in the direction taken for $\underline{n}$, the solid angle changes discontinuously by -4π. This property is true for any surface not necessarily closed.

For, let S_1 be the surface. Add another surface S_2 such that $S_1 + S_2$ makes a simple closed surface S. Let w_1, w_2, w be the solid angles subtended by S_1, S_2, S respectively,

at any point O. If O crosses S_1 in the direction of the outward normal, w decreases from 4π to 0, that is, it is discontinuous by -4π. But the solid angle ω_2 is clearly continuous as O crosses S_1. Hence, since $w = w_1 + w_2$, it follows that w_1 is discontinuous by -4π as O crosses the surface.

13. Review of Gauss-Green's theorem

13.1.1972

Deductions from the divergence theorem of Gauss - Green's theorem.

$$\iiint_V (\nabla \cdot \vec{F})\, dV = \oiint_S F \cdot n\, dS$$

Put $\bar{F} = \phi \nabla \psi$.

Where ϕ & ψ are two scalar point fns of (x, y, z) assumed continious & differential.

$$\iiint_V [\nabla \cdot (\phi \nabla \psi)]\, dV = \oiint_S \phi \nabla \psi \cdot \bar{n}\, dS.$$

Now we have already shown that

$$\operatorname{div}(\phi \bar{A}) = \phi \operatorname{div} \bar{A} + \bar{A} \cdot \nabla \phi$$

put $\bar{A} = \nabla \psi$.

$$\therefore \nabla\cdot(\phi\nabla\psi) = \phi(\nabla\cdot\nabla\psi) + \nabla\psi\cdot\nabla\phi.$$

$$\therefore \iiint_V (\phi\nabla^2\psi + \nabla\psi\cdot\nabla\phi)\,dV = \oiint_S \phi\nabla\psi\,\hat{n}\,ds \quad \rightarrow (1)$$

This is known as the first form of Green's theorem.

$$\iiint_V (\phi\nabla^2\psi + \nabla\psi\cdot\nabla\phi)\,dV = \oiint_S \phi\frac{\partial\psi}{\partial n}\,dS$$

$\nabla\psi$

Since the scalar component of the gradient of the scalar fn in any direction equal to the rate of change of scalar fn in that direction. now interchange ψ & ϕ

$$\iiint_V (\psi\nabla^2\phi + \nabla\phi\cdot\nabla\psi)\,dV = \oiint_S \psi\frac{\partial\phi}{\partial n}\,dS$$

Substracting :-

$$\iiint_V (\phi \nabla^2 \psi - \psi \nabla^2 \phi)\, dv = \iint_S \left(\phi \frac{\partial \psi}{\partial n} - \psi \frac{\partial \phi}{\partial n}\right) ds$$

which is known as the second form or symmetrical form of green's theory.

Let us consider the particular case $\psi = 1$

$$\therefore \iiint_V (\nabla^2 \phi)\, dv = \oiint_S \frac{\partial \phi}{\partial n}\, dS$$

triple double

ex.

A vector field is given by $\bar{A} = 3x^2 y \bar{i} + x y^2 \bar{j}$. evaluate the surface integral $\oiint_S \bar{A} \cdot \bar{n}\, ds$ over a closed surface of a triangular prism of unit hieght in the z-direction whose base is bounded by the positive x-y axis & the line $x + y = 1$. And verify the divergence theorem of gauss in this case

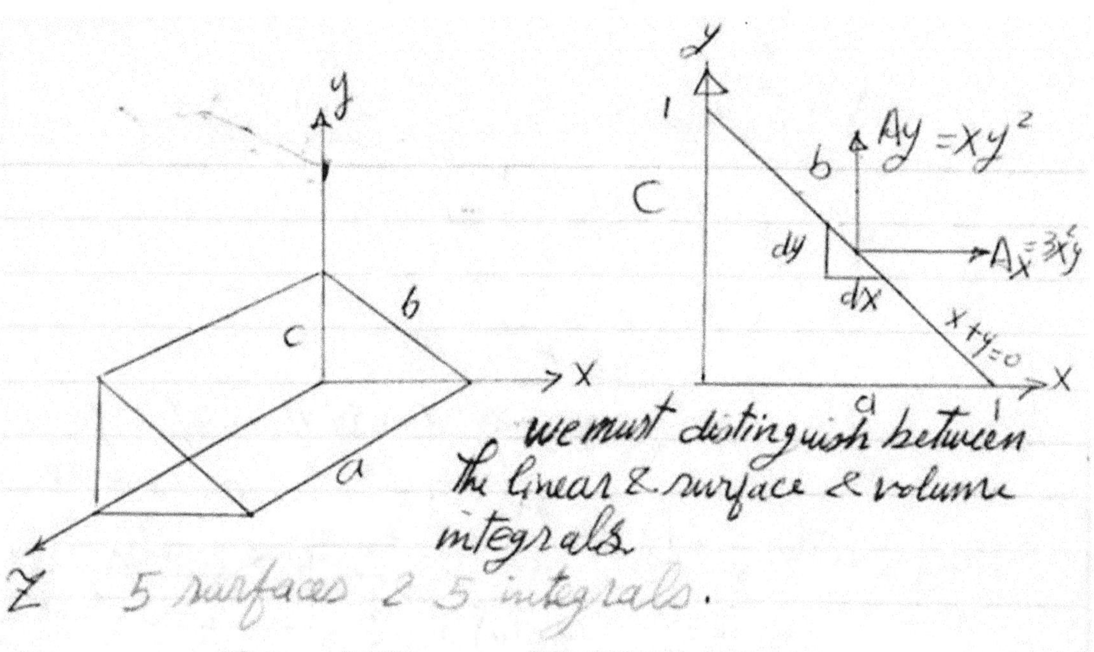

we must distinguish between the linear & surface & volume integrals.

5 surfaces & 5 integrals.

Sol$^{\underline{n}}$

Since the vector "A" has no component in the Z direction then the surface integral of A over the top & botton of prism = 0

Next consider the surface integral over the surface "a"

$-\int_0^1 xy^2 dx$ with $y = 0$ —— ① = 0

because the positive direction is Convergence or outward "a" ↓

y, A_y, xy^2, $3x^2y$, c, a, x, xy^2, divergence = −ve

Next consider the surface integral of "A" over the surface "C"

$$-\int_0^1 3x^2 y\,dy \text{ with } x=0 \quad = 0 \rightarrow (2)$$

$\therefore$ 4 integrals vanished $= 0$

Now consider the surface integral over the surface "b"

$$\int_b A_x dy - A_y dx$$

$A_y = xy^2$

$A_x = 3x^2y$

+ve direct^n

$$-\int A_y dx - (A_x dy)$$
$$= \int A_x dy - A_y dx$$

$$= \int_b 3x^2 y\,dy - xy^2 dx$$

as $x + y = 1 \quad \therefore \quad x = 1 - y$

$dx = -dy$

$$\therefore \int_b A_x dy - A_y dx = \int_0^1 3(1-y)^2 y + (1-y)y^2 dy$$

$$= \frac{1}{3}$$

$$\therefore \iint_S \vec{A}\,\bar{n}\,ds = \frac{1}{3} \longrightarrow (1)$$

$$\nabla\cdot\bar{A} = \frac{\partial}{\partial x}(3x^2y) + \frac{\partial}{\partial y}(xy^2)$$

$$= 6xy + 2xy = 8xy$$

$$\iiint_V (\nabla\cdot\bar{A})\,dv = \iint_S 8xy\,dx\,dy$$

$$= \int_0^1\int_0^{1-x} 8xy\,dx\,dy$$

c, b, $x+y=1$, (i), (ii), a, 1

$$= \int_0^1 \left[\frac{8}{2}xy^2\right]_0^{1-x} dx$$

$$= 4\int_0^1 x(1-x)^2\,dx = \frac{1}{3}$$

we analysis the vector perpendicular to the surface outwards & ...

14. Examples on Green's Theorem

1. Write the divergence theorem in its rectangular form.

The divergence theorem is given by

$$\int_{\tau} \nabla \cdot \underline{A}\, d\tau = \int_{s} \underline{A}\cdot\underline{n}\, dS$$

Let $\underline{A} = A_1\underline{i} + A_2\underline{j} + A_3\underline{K}$, $\nabla\cdot\underline{A} = \frac{\partial A_1}{\partial x} + \frac{\partial A_2}{\partial y} + \frac{\partial A_3}{\partial z}$

$\underline{n} = n_1\underline{i} + n_2\underline{j} + n_3\underline{k}$. Then $n_1 = \underline{n}.\underline{i} = \cos\alpha$, $n_2 = \underline{n}.\underline{j} = \cos\beta$

$n_3 = \underline{n}.\underline{k} = \cos\gamma$.

$\therefore\ \underline{A}\cdot\underline{n} = A_1 \cos\alpha + A_2 \cos\beta + A_3 \cos\gamma$

$$\therefore \iiint_{\tau} \left(\frac{\partial A_1}{\partial x} + \frac{\partial A_2}{\partial y} + \frac{\partial A_3}{\partial z}\right) dx\, dy\, dz$$

$$= \int_{s} (A_1 \cos\alpha + A_2 \cos\beta + A_3 \cos\gamma)\, dS$$

2. Evaluate $\int_{s} \underline{r}.\underline{n}\, dS$, where S is a closed surface.

By the divergence theorem,

$$\int_{s} \underline{r}.\underline{n}\, dS = \int_{\tau} \nabla\cdot\underline{r}\, d\tau$$

$$= \int_{\tau} \left(\frac{\partial}{\partial x}\underline{i} + \frac{\partial}{\partial y}\underline{j} + \frac{\partial}{\partial z}\underline{k}\right).(x\underline{i}+y\underline{j}+z\underline{k})\, d\tau$$

$$= \int_{\tau} \left(\frac{\partial x}{\partial x} + \frac{\partial y}{\partial y} + \frac{\partial z}{\partial z}\right) dV = 3\int d\tau$$

$$= 3V$$

where V is the volume enclosed by S.

3. Show that $\int_\tau \underline{r}.\text{curl } \underline{r}\, d\tau = 0$, where $\underline{r} = x\underline{i} + y\underline{j} + z\underline{k}$ and the region of integration τ is the interior of the sphere $r = a$.

Since curl $\underline{r} = 0$, then

$$\text{div } \underline{V} \wedge \underline{r} = \underline{r}.\text{curl } \underline{V} - \underline{V}.\text{ curl } \underline{r}$$
$$= \underline{r}.\text{ curl } \underline{V}.$$

$$\int_\tau \underline{r}.\text{ curl } \underline{V}\, d\tau = \int_\tau \text{div } (\underline{V} \wedge \underline{r}) d\tau$$
$$= \int_\tau \underline{n}.\ (\underline{V} \wedge \underline{r}) d\tau$$
$$= \int_{r=a} \underline{V}.(\underline{n} \wedge \underline{r}) dS$$

On the surface of the sphere $\underline{n}$ and $\underline{r}$ are in the same direction. Therefore $\underline{n} \wedge \underline{r} = 0$ and

$$\int_\tau \underline{r}.\text{curl } \underline{V}\, d\tau = 0.$$

4. Verify the divergence theorem for $\underline{A} = 4x\underline{i} - 2y^2\underline{j} + z^2\underline{k}$ taken over the region bounded by $x^2 + y^2 = 4$, $z = 0$ and $z = 3$.

$$\text{Volume integral} = \int_\tau \nabla.\underline{A}\, d\tau$$

$$= \int_\tau \left[\frac{\partial}{\partial x}(4x) + \frac{\partial}{\partial y}(-2y^2) + \frac{\partial}{\partial z}(z^2)\right] d\tau$$

$$= \int_\tau (4-4y+2z)\,d\tau$$

$$= \int_{-2}^{2} \int_{-\sqrt{4-x^2}}^{\sqrt{4-x^2}} \int_{0}^{3} (4-4y+2z)\,dz\,dy\,dx$$

$$= 84\pi .$$

The surface S of the cylinder consists of the base $S_1(z=0)$, the top $S_2(z=3)$ and the convex portion $S_3(x^2+y^2=4)$.

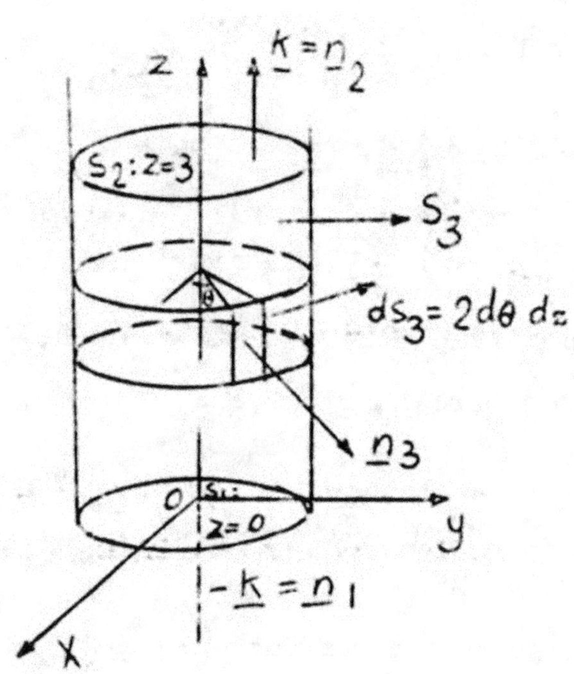

$\therefore$ the surface integral

$$= \int_S \underline{A}.n\, dS = \int_{S_1} \underline{A}.\underline{n}\, dS_1 + \int_{S_2} \underline{A}.\underline{n}\, dS_2 + \int_{S_3} \underline{A}.\underline{n}\, dS_3$$

On $S_1(z=0)$, $\underline{n} = -\underline{k}$, $\underline{A} = 4x\underline{i} - 2y^2\underline{j}$ and $\underline{A}.\underline{n} = 0$, so that

$$\int_{S_1} (\underline{A}.\underline{n})dS_1 = 0$$

On $S_2(z = 3)$, $\underline{n} = \underline{k}$, $\underline{A} = 4x\underline{i} - 2y^2\underline{j} + 9\underline{k}$ and $\underline{A}.\underline{n} = 9$,

So that

$$\int_{S_2} (\underline{A}.\underline{n})dS_2 = 9 \int_{S_2} dS_2 = 36\pi$$, since the area of $S_2 = 4\pi$.

On $S_3(x^2 + y^2 = 4)$. A perpendicular to $x^2 + y^2 = 4$ has the direction of $\nabla(x^2 + y^2) = 2x\underline{i} + 2y\underline{j}$.

Then the unit normal $\underline{n} = \dfrac{2x\underline{i} + 2y\underline{j}}{\sqrt{4x^2 + 4y^2}} = \dfrac{x\underline{i} + y\underline{j}}{2}$,

$\underline{A}.\underline{n} = (4x\underline{i} - 2y^2\underline{j} + z^2\underline{k}).(\dfrac{x\underline{i} + y\underline{j}}{2}) = 2x^2 - y^2$

From the figure, $x = 2\cos\theta$, $y = 2\sin\theta$, $dS_3 = 2\, d\theta\, dz$

and so,

$$\int_{S_3} \underline{A}.\underline{n}\, dS_3 = \int_0^{2\pi}\int_0^3 \left[2(2\cos\theta)^2 - (2\sin\theta)^3\right] 2\, dz\, d\theta$$

$$= \int_0^{2\pi} (48 \cos^2\theta - 48 \sin^3\theta) d\theta = \int_0^{2\pi} 48 \cos^2\theta \, d\theta$$

$$= 48\pi$$

Then the surface integral $= 36\pi + 48\pi = 84\pi$, agreeing with the volume integral and verifying the divergence theorem.

5. Determine the function ϕ as a function of position inside a sphere of unit radius given that $\nabla^2\phi = 6$ everywhere inside the sphere and that everywhere on the boundary of the sphere $\phi = 0$ and $\frac{\partial \phi}{\partial n} = 2$.

Substitute in Green's formula,

$$4\pi\phi = \int_S \left(\frac{1}{r}\frac{\partial \phi}{\partial n} - \phi \frac{\partial}{\partial n}\frac{1}{r}\right) dS - \int \frac{1}{r} \nabla^2\phi \, d\tau,$$

we get for a point at a distance R from the centre of the sphere

$$4\pi\phi = 2 \int_S \frac{dS}{r} - 6 \int_\tau \frac{d\tau}{r} .$$

Carrying out the integration we obtain

$$4\pi\phi = 8\pi - 6\pi\left(2 - \frac{2}{3}R^2\right)$$

i.e. $\phi = R^2 - 1.$

6. Show that the area bounded by a simple closed curve C is given by $\frac{1}{2}\oint_C xdy - ydx$.

In Green's theorem, put Q = -y, P = x, then:

$$\oint_C xdy - ydx = \iint_R \left[\frac{\partial}{\partial x}(x) - \frac{\partial}{\partial y}(-y)\right] dx\, dy$$

$$= 2\iint_R dx\, dy = 2A,$$

where A is the required area. Thus $A = \frac{1}{2}\oint_C xdy - ydx$.

7. Evaluate $\oint_C (y-\sin x)dx + \cos x\, dy$, where C is the triangle of the adjoining figure:

(a) directly,

(b) by using Green's theorem in the plane.

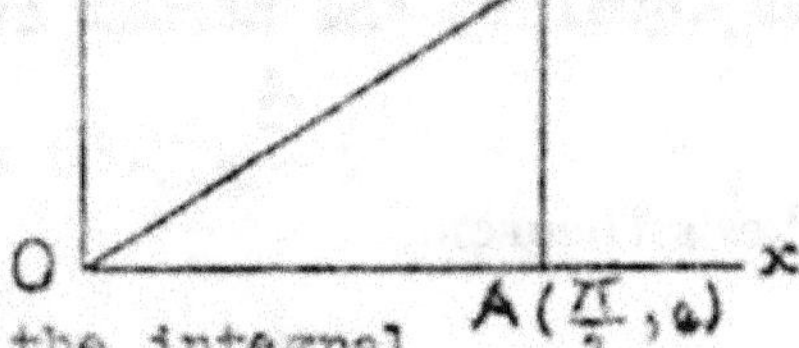

(a) Along OA, y = 0, dy = 0 and the integral

$$\int_0^{\frac{\pi}{2}} -\sin x\, dx = -1 .$$

Along AB, $x = \frac{\pi}{2}$, dx = 0 and the integral equals zero.

Along BO, $y = \frac{2x}{\pi}$, $dy = \frac{2}{\pi} dx$ and the integral equals

$$\int_{\frac{\pi}{2}}^{0} \left(\frac{2x}{\pi} - \sin x\right)dx + \frac{2}{\pi} \cos x \, dx = 1 - \frac{\pi}{4} - \frac{2}{\pi}$$

Then the integral along $C = -\frac{\pi}{4} - \frac{2}{\pi}$.

(b) $Q = y - \sin x$, $P = \cos x$, $\frac{\partial P}{\partial x} = -\sin x$, $\frac{\partial Q}{\partial y} = 1$ and

$$\oint_C Q\,dx + P\,dy = \iint_R \left(\frac{\partial P}{\partial x} - \frac{\partial Q}{\partial y}\right)dx\,dy$$

$$= \int_0^{\frac{\pi}{2}} \int_0^{\frac{2x}{\pi}} (-\sin x - 1)\,dy\,dx = -\frac{2}{\pi} - \frac{\pi}{4}$$

in agreement with part (a).

15. Stokes's Theorem

Consider an open surface S bounded by a closed non-intersecting curve C. Let $\underline{n}$ be the unit vector normal to S in the direction to which the head of a man points when he walks along C in the sense it is described while keeping S to his left. Then for a vector point-function $\underline{V}$ which is single-valued and continuous along with its derivative in any direction, the tangential line integral of $\underline{V}$ round the curve C is equal to the normal surface integral of curl $\underline{V}$ over S; that is

$$\oint_C V.dr = \int_S \text{curl } \underline{V}.\underline{n}\,dS .$$

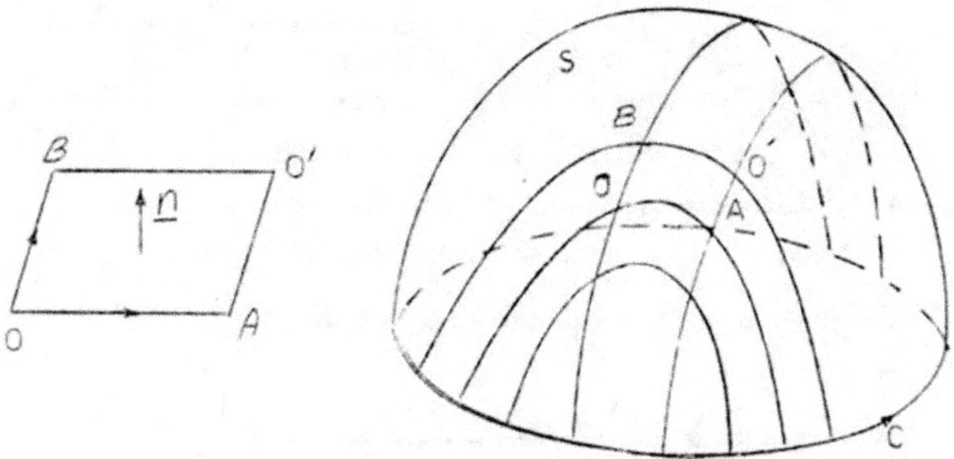

Divide the surface S into infinitesimal meshes by a set of curves joining points of C. Consider one of these meshes OAO'B. Without loss of generality, the typical mesh can be taken in the form of an infinitesimal parallelogram. Let its sides OA & OB

16. Theorem deducible from Stokes 's theorem

represent the infinitesimal vectors $\underline{a}$, $\underline{b}$ respectively.

Let $\underline{V}$ be the value of the vector point function at the centre of the element. Its value at the middle of OB

$$= \underline{V} - (\frac{\underline{a}}{2} . \nabla) \underline{V},$$

and at the middle of AO' $= \underline{V} + (\frac{\underline{a}}{2} . \nabla) \underline{V}$.

The values at the middle of OA and BO' are:

$$\underline{V} - \tfrac{1}{2}(\underline{b} \cdot \nabla)\underline{V} \text{ and } \underline{V} + \tfrac{1}{2}(\underline{b} \cdot \nabla)\underline{V},$$

respectively. Since these edges are infinitesimal, these values may be taken to be the value of the vector function over the respective sides.

$$\therefore \int_{OAO'B} \underline{V}.d\underline{r} = \left(\int_{OA} \underline{V}.d\underline{r} - \int_{BO'} \underline{V}.d\underline{r}\right) + \left(\int_{AO'} \underline{V}.d\underline{r} - \int_{BO} \underline{V}.d\underline{r}\right)$$

$$= -\left\{(\underline{b}.\nabla)\underline{V}\right\}.\underline{a} + \left\{(\underline{a}.\nabla)\underline{V}\right\}.\underline{b}$$

$$= \left\{\underline{b}(\underline{a}.\nabla) - \underline{a}(\underline{b}.\nabla)\right\}.\underline{V}$$

$$= \left\{(\underline{a}\wedge\underline{b})\wedge\nabla\right\}.\underline{V}$$

$$= (\underline{a}\wedge\underline{b}).\nabla\wedge\underline{V}.$$

But $\underline{a}\wedge\underline{b} = \underline{n}\, dS$; therefore

$$\int_{OAO'B} \underline{V}.d\underline{r} = \underline{n}.\nabla\wedge\underline{V}\, dS$$

Add for the different meshes. For two neighbouring meshes, the integral taken over the common bounding sides will cancel. This is true so long as $\underline{V}$ and its first derivatives are single-valued and continuous. Adding for all meshes, we will be left

17. Stokes's theorem for a plane region

with $\oint_C \underline{V}.d\underline{r}$. In doing so, we are at the same time integrating $\underline{n}.\nabla\wedge\underline{V}$ over the whole surface S.

$$\therefore \oint_C \underline{V}.d\underline{r} = \int_S \underline{n}.\nabla\wedge\underline{V}\, dS$$

Notice that the way we proved the theorem, shows that it can be applied to surfaces S bounded by more than one curve, as the one shown in figure, in which case the theorem takes the form:

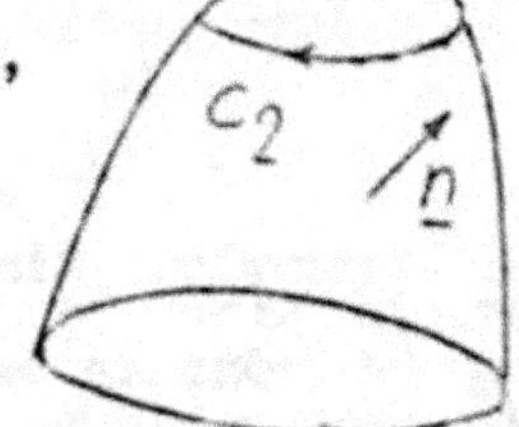

$$\int \underline{n}.(\nabla\wedge\underline{V})\, dS = \oint_{C_1} \underline{V}.d\underline{r} + \oint_{C_2} \underline{V}.d\underline{r}$$

Applying Stokes's theorem to the vector point-function $\underline{d}\ \underline{V}$ where $\underline{d}$ is a constant vector, we get

$$\oint_C (\underline{d}\wedge\underline{V}).d\underline{r} = \int_S \underline{n}.\nabla\wedge(\underline{d}\wedge\underline{V})\, dS$$

$$\oint_C \underline{d}.(\underline{V}\wedge d\underline{r}) = -\int_S (\underline{n}\wedge\nabla).\underline{V}\wedge\underline{d}\, dS$$

$$= -\int_S (\underline{n}\wedge\nabla)\wedge\underline{V}\Big\}.\,\underline{d}\, dS$$

$$\therefore\ \underline{d}.\oint_C d\underline{r}\wedge\underline{V} = \underline{d}.\int_S (\underline{n}\wedge\nabla)\wedge\underline{V}\, dS$$

Since this is true for any constant vector $\underline{d}$, we must have

$$\int_C d\,\underline{r}\wedge\underline{V} = \int_S (\underline{n}\wedge\nabla)\wedge\underline{V}\, dS\ .$$

Also by applying Stokes's theorem to $\underline{d}\ \emptyset$ where $\emptyset$ is a scalar point function we get:

$$d . \oint_C \phi \, d\underline{r} = \int_S (\nabla \wedge \underline{d}\phi) . \underline{n} \, dS$$

$$= d . \int_S \underline{n} \wedge \nabla \phi \, dS$$

i.e. $$\oint_C \phi \, d\underline{r} = \int_S \underline{n} \wedge \nabla \phi \, dS .$$

Let the area S enclosed by the simple curve C lie in the x,y plane and let P, Q and their derivatives be single-valued and continuous point functions.

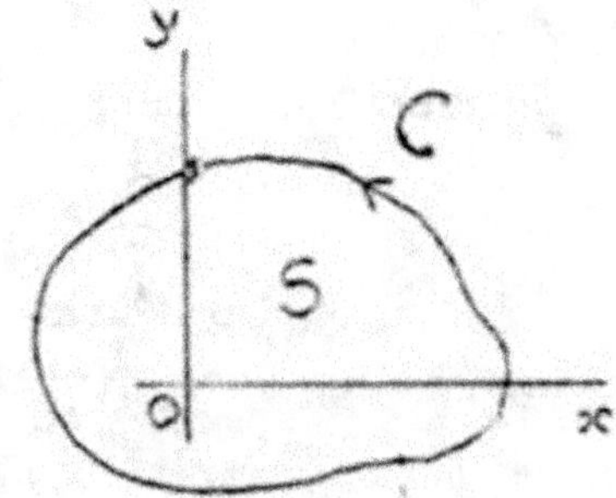

As $\underline{n} = \underline{k}$, and $d\underline{r} = \underline{i}\, dx + \underline{j}\, dy$, Stokes's theorem applied to $\underline{V} = \underline{i}\, P + \underline{j}\, Q$ gives

$$\oint_C (P\, dx + Q\, dy) = \int_S \left(\frac{\partial Q}{\partial x} - \frac{\partial P}{\partial y}\right) dx\, dy$$

which is the same as Green's theorem in the plane.

18. Examples on Stokes's Theorem

1. If $\oint_C \underline{E}.d\underline{r} = -\frac{1}{c}\frac{\partial}{\partial t}\int_S \underline{H}.d\underline{S}$ where S is any surface bounded by the curve C, show that,

$$\nabla \wedge \underline{E} = -\frac{1}{c}\frac{\partial \underline{H}}{\partial t}.$$

From Stokes's theorem, we have

$$\int_S (\nabla \wedge \underline{E}).dS = \oint_C \underline{E}.d\underline{r} = -\frac{1}{c}\frac{\partial}{\partial t}\int_S \underline{H}.d\underline{S}.$$

$$\text{i.e.} \quad \int_S (\nabla \wedge \underline{E}).d\underline{S} = -\frac{1}{c}\int_S \frac{\partial \underline{H}}{\partial t}.d\underline{S}$$

Since S is arbitrary, we get

$$\nabla \wedge \underline{E} = -\frac{1}{c}\frac{\partial \underline{H}}{\partial t}.$$

2. Prove that $\oint_C (\underline{a} \wedge \underline{r}).d\underline{r} = 2\underline{a}.\int_S \underline{n}\, dS$, where $\underline{a}$ is any constant vector and the line integral is taken round the closed boundary of the surface S. Hence show that

$$\oint_C \underline{r} \wedge d\underline{r} = 2\int_S \underline{n}\, dS .$$

Applying Stokes's theorem to the vector $\underline{a} \wedge \underline{r}$, we get,

$$\oint_C (\underline{a} \wedge \underline{r}).d\underline{r} = \int_S (\nabla \wedge (\underline{a} \wedge \underline{r})).\underline{n}\, dS .$$

But for the constant vector a, $\nabla \wedge (\underline{a} \wedge \underline{r}) = 2\underline{a}$, hence

$$\oint_C (\underline{a} \wedge r).d\underline{r} = 2\underline{a}. \int_S \underline{n}\, dS .$$

i.e. $$\underline{a} \oint_C \underline{r} \wedge d\underline{r} = 2\underline{a}. \int_S \underline{n}\, dS.$$

Since $\underline{a}$ is arbitrary, we have

$$\oint_C \underline{r} \wedge d\underline{r} = 2\int_S \underline{n}\, dS.$$

3. Verify Stokes's theorem for $\underline{A} = (2x - y)\underline{i} - yz^2\underline{j} - y^2z\underline{k}$, where S is the upper half surface of the sphere $x^2 + y^2 + z^2 = 1$; and C is its boundary.

The boundary of S is a circle in the xy plane of radius one and centre at the origin.
Let $x = \cos t$, $y = \sin t$, $z = 0$, $0 \leq t < 2\pi$ be the parametric equations of C.

$$\oint_C \underline{A}.d\underline{r} = \oint_C (2x - y)dx - yz^2dy - y^2z\, dz$$

$$= \int_0^{2\pi} (2 \cos t - \sin t)(-\sin t)dt = \pi.$$

Also, $$\nabla \wedge \underline{A} = \begin{vmatrix} \underline{i} & \underline{j} & \underline{k} \\ \frac{\partial}{\partial x} & \frac{\partial}{\partial y} & \frac{\partial}{\partial z} \\ 2x-y & -yz^2 & -y^2z \end{vmatrix} = \underline{k}$$

Then

$$\int_S (\nabla \wedge \underline{A}).\underline{n}\, dS = \int_S (\underline{k}.\underline{n})\, dS = \iint_R dx\, dy ,$$

since $\underline{n}.\underline{k}\, dS = dx\, dy$ and R is the projection of S on the xy-plane. This last integral equals :

$$\int_{-1}^{1}\int_{-\sqrt{1-x^2}}^{\sqrt{1-x^2}} dy\, dx = 4\int_{0}^{1}\int_{0}^{\sqrt{1-x^2}} dy\, dx = 4\int_{0}^{1} \sqrt{1-x^2}\, dx = \pi.$$

and Stoke's theorem is verified.

4. Prove that a necessary and sufficient condition that: $\oint_C \underline{A}.d\underline{r} = 0$ for every closed curve C is that $\nabla\wedge\underline{A} = 0$ identically.

Sufficiency: Suppose $\nabla\wedge\underline{A} = 0$. Then by Stoke's theorem,

$$\oint_C \underline{A}.d\underline{r} = \int_S (\nabla\wedge\underline{A}).\underline{n}\, dS = 0$$

Necessity: Suppose $\oint_C \underline{A}.d\underline{r} = 0$ around every closed path C, and assume $\nabla\wedge\underline{A} \neq 0$ at some point P. Then assuming $\nabla\wedge\underline{A}$ is continuous, there will be a region with P as an interior point, where $\nabla\wedge\underline{A} \neq 0$. Let S be a surface contained in this region whose normal $\underline{n}$ at each point has the same direction as $\nabla\wedge\underline{A}$, i.e. $\nabla\wedge\underline{A} = \alpha\underline{n}$ where α is a positive constant.

Let C be the boundary of S. Then by Stoke's theorem:

$$\oint_C \underline{A}.d\underline{r} = \iint_S (\nabla\wedge\underline{A}).\underline{n}\, dS = \alpha\int_S \underline{n}.\underline{n}\, dS > 0$$

which contradicts the hypothesis that $\oint_C \underline{A}.d\underline{r} = 0$ and shows that $\nabla\wedge\underline{A} = 0$.

It follows that $\nabla\wedge\underline{A}=0$ is also a necessary and sufficient condition for a line integral $\int_{P_1}^{P_2} \underline{A}.d\underline{r}$ to be independent of the path joining points P_1 and P_2.

19. Review of Stokes' Theorem

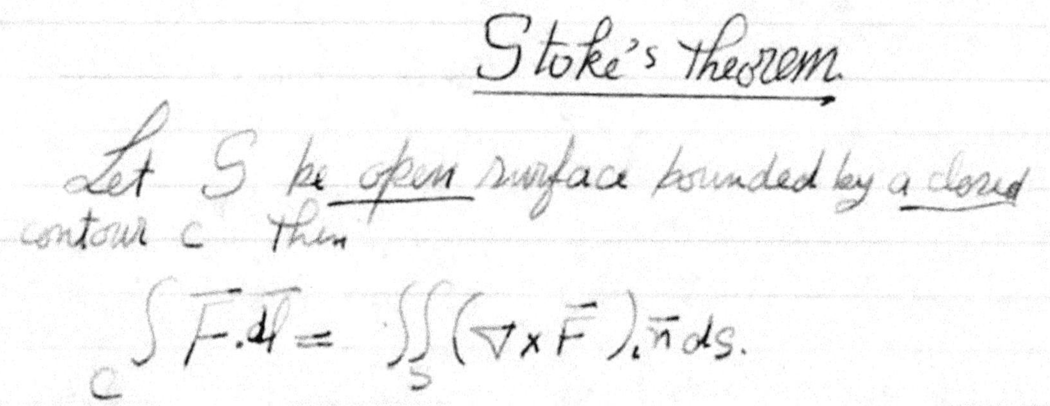

Stoke's theorem

Let S be open surface bounded by a closed contour C then

$$\int_C \bar{F}.\bar{dl} = \iint_S (\nabla \times \bar{F}).\bar{n}\,ds.$$

i.e The line integral of F along the boundary C is equal to the surface integral of curl F over the surface S

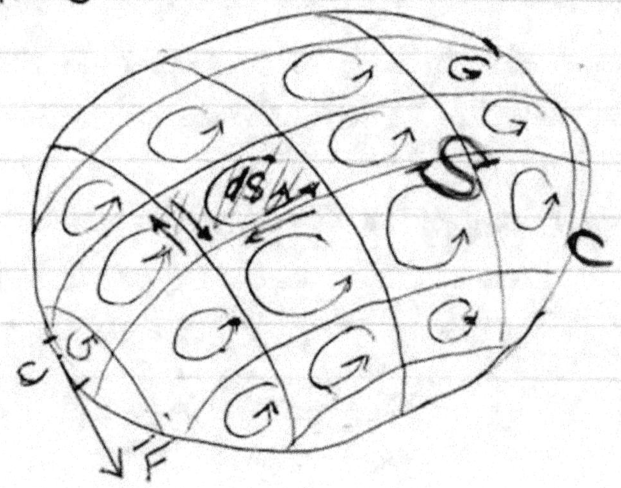

Divide the area S into a large small areas by two sets of intersected curves. & consider one of these areas

We have already shown that the line integral of F along the boundary of the area ds divided by the area is equal in the limit to the scalar compt of curl F normal to the area.

i.e. the line integral of $\bar{F}$ along the boundary
= the scalar component of curl F. normal to the area multiplied by the area.

$$\text{as} \lim_{\text{area}\to 0} \frac{\int_c \bar{F}\cdot d\bar{l}}{\text{area}} = \text{curl } F$$

$$\therefore \frac{d}{d\,\text{area}} \int_c \bar{F}\cdot d\bar{l} = \text{curl}.F$$

$$\therefore \int_c \bar{F}\cdot d\bar{l} = \text{curl} F \cdot \text{area}.$$

Now the integral Along AB inters twice with opposite sides & hence they cancel one another. similarly for all signals except those along the Boundary whose sum is

$\int_C F\, dl$ This is equal to the surface integral of curl F over the the surface S. & hence

$$\int_C \bar{F} \cdot d\bar{l} = \iint_S (\nabla \times \bar{F})\, \bar{n}\, ds$$

open therefore there is not $\oint$

ex.

$\bar{F} = (x^2+y^2)\,\bar{i} - 2xy\,\bar{j}$ evaluate the surface integral $\iint_S (\nabla \times \bar{F})\, \bar{n}\, ds$ over rectangular area in the x-y plane Bouded by the lines $x=0$, $x=a$, $y=0$, $y=b$ & verify stokes theorem.

Sol^n

$$\nabla \times \bar{F} = \begin{vmatrix} i & j & k \\ \frac{\partial}{\partial x} & \frac{\partial}{\partial y} & \frac{\partial}{\partial z} \\ x^2+y^2 & -2xy & 0 \end{vmatrix}$$

$$= (-2y-2y)k = -4y\bar{k}$$

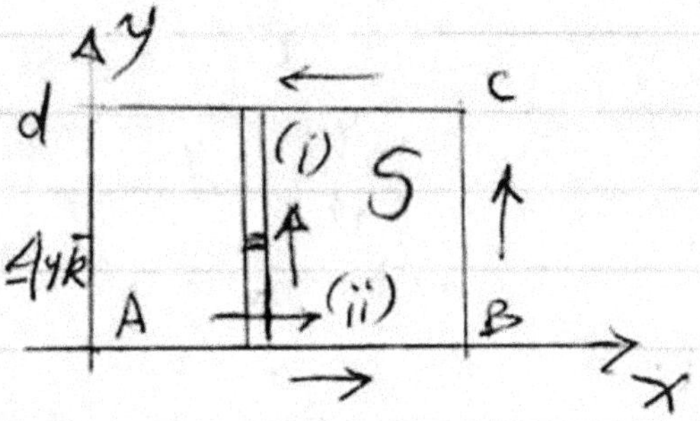

$\bar{k} = \bar{n}$

$\therefore \bar{k}.\bar{n} = 1$

$$\iint_S (\nabla \times \bar{F}).\bar{n}\, ds = \iint_S -4y\, dx dy$$

$$= \int_0^a \int_0^b -4y\, dx\, dy$$

$$= \int_0^a \left(-2y^2\right)_0^b = -2b^2 a$$

Now we try to prove that the surface integral $-2b^2a$ = line integral of $\int \bar{F}.d\bar{l}$

$$\oint_{c} F.d\bar{l} = \int_A^B + \int_B^C + \int_C^D + \int_D^A$$

closed contour

1st. $\int_A^B \bar{F}\cdot\bar{dl} = \int_0^a (x^2+y^2)\,dx$ with $y=0$

$= \frac{a^3}{3}$ —(1)

2nd $\int_B^c \bar{F}\cdot\bar{dl} = \int_0^b -2xy\,dy$ with $x=a$

$= -2a\frac{b^2}{2} = -ab^2$ —(2)

3nd $\int_C^d \bar{F}\cdot\bar{dl} = -\int_0^a (x^2+y^2)\,dx$ with $y=b$

$= -\left(\frac{a^3}{3}+b^2a-0\right) = -\frac{a^3}{3}-b^2a$ —(3)

4th $\int_d^A \bar{F}\cdot\bar{dl} = -\int_0^b -2xy\,dy$ with $x=0$

$= 0$

Adding, $\oint_{ABCd} \bar{F}\cdot\bar{dl} = -2b^2a.$

20. Exercises on Stokes' theorem

1. If $\underline{V} = \nabla \phi$ and $\nabla^2 \phi = -4\pi\rho$, show that for a closed surface S bounding a volume τ

$$\int_S \underline{V}.\underline{n}\, dS = -4\pi \int_\tau \rho\, d\tau,$$

$$\int_S \phi\, \underline{V}.\underline{n}\, dS = \int_\tau V^2\, d\tau - 4\pi \int_\tau \rho\,\phi\, d\tau.$$

2. If $c\rho \frac{\partial V}{\partial t} = \text{div}\,(k \nabla V)$ and V is zero over the closed surface S, show that

$$\int c\rho V \frac{\partial V}{\partial t}\, d\tau = -\int k(\nabla V)^2 d\tau.$$

3. If $\underline{W} = \frac{1}{2}$ curl $\underline{V}$ and $\underline{V}$ = curl $\underline{U}$, show that

$$\frac{1}{2}\int V^2 d\tau = \frac{1}{2}\int (\underline{U} \wedge \underline{V}).\underline{n}\, dS + \int \underline{U}.\underline{W}\, d\tau.$$

4. For a closed surface S and point functions $\underline{V}$, ϕ, prove that the integrals

$$\int_S \underline{n}.\,\text{curl}\,\underline{V}\, dS \quad \text{and} \quad \int_S \underline{n} \wedge \text{grad}\,\phi\, dS$$

vanish identically.

5. Show that:

$$\int_\tau \underline{V}.\nabla p\, d\tau = \int_S p\underline{V}.\underline{n}\, dS - \int_\tau p\, \text{div}.\underline{V}\, d\tau$$

and

$$\int_\tau \underline{V}.\,\text{curl}\,\underline{U}\, d\tau = \int_S \underline{U} \wedge \underline{V}.\underline{n}\, dS + \int_\tau \underline{U}.\,\text{curl}\,\underline{V}\, d\tau$$

6. If $\underline{n}$ is the unit outward drawn normal to any surface of area S, show that $\int_\tau \text{div}.\underline{n}\, d\tau = S$.

7. Prove

a) $\int_\tau \frac{d\tau}{r^2} = \int_S \frac{\underline{r}.\underline{n}}{r^2}\, dS$,

b) $\int_\tau 5r^3\, \underline{r}\, d\tau = \int_S r^5\, \underline{n}\, dS.$

8. If $\underline{U}$ and $\underline{V}$ are vector point functions of a scalar parameter t and are connected by the relations

$$\frac{\partial \underline{U}}{\partial t} = \text{curl } \underline{V}\ , \quad \frac{\partial \underline{V}}{\partial t} = -\text{ curl } \underline{U},$$

and if τ is the volume enclosed by a fixed surface S, show

that

$$\frac{\partial}{\partial t} \int_\tau (\underline{U}^2 + \underline{V}^2)\, d\tau = 2\int (\underline{V} \wedge \underline{U}).\underline{n}\, dS$$

where $\underline{n}$ is the outward unit vector normal to S.

9. Determine the function ϕ as a function of position inside a sphere of unit radius given that $\nabla^2 \phi = 12$ everywhere inside the sphere and that everywhere on the boundary of the sphere $\phi = 0$ and $\frac{\partial \phi}{\partial n} = 4.$

10. A solution of the wave equation

$$\frac{\partial^2 \psi}{\partial t^2} = c^2 \nabla^2 \psi,$$

is of the form $\psi = \phi(r)\, f(t)$. Prove that if f(t) is periodic, then ϕ satisfies the equation,

$$\nabla^2 \phi = -\, k^2 \phi\, ,$$

where k is a constant.

Show that $\frac{e^{ikr}}{r}$ is a solution of this equation in any region not including the origin, and use Green's theorem to prove that for every solution, the value of ϕ at the origin is given by

$$-\frac{1}{4\pi}\int_S \left[\phi \frac{\partial}{\partial n} \frac{e^{ikr}}{r} - \frac{e^{ikr}}{r} \frac{\partial \phi}{\partial n} \right] dS$$

assuming that the origin is within S.

11. Deduce the relation,

$$\int \phi \underline{n}.\text{curl } \underline{V} \, dS = \int \phi \underline{V}.\, d\underline{r} - \int \nabla \phi \wedge \underline{V}.\underline{n} \, dS .$$

Putting $\underline{V} = \nabla \psi$ in this result, prove that

$$\int (\nabla \phi \wedge \nabla \psi).\underline{n} \, dS = \int \phi \nabla \psi . d\underline{r} = - \int \psi \nabla \phi . d\underline{r}.$$

12. Prove that div $z\underline{r} = 4z$, where $\underline{r} = \underline{i}x + \underline{j}y + \underline{k}z$. Hence, use Gauss' theorem to evaluate $\int_\tau z dt$, where τ is the volume bounded by the x,y-plane and $x^2 + y^2 + z^2 = a^2$, $z > 0$.

13. If $\underline{F} = (y^2 + z^2 - x^2,\ z^2 + x^2 - y^2,\ x^2 + y^2)$, determine the fluxes of $\underline{F}$ and curl $\underline{F}$ across the curved surface of the hemisphere $x^2 + y^2 + z^2 - 2ax = 0$, $z \geq 0$.

14. Calculate the surface integral $\int_S \underline{a}.\underline{n}\, dS$ of the vector $\underline{a}$ $(xy, 2yz, 0)$ over the surface of the tetrahedron bounded by the planes $x = 0$, $y = 0$, $z = 0$ and $x + y + 2z = 2$. Hence find the volume integral $\int (2y + 4z)d\tau$ over the tetrahedron.

15. Verify Green's theorem in the plane for

$$\oint_C (xy + y^2)dx + x^2 dy,$$

where C is the closed curve whose equations are $x = y$ and $y = x^2$.

16. Given the vector $\underline{A} = (x^2 - y^2)\underline{i} + 2xy\,\underline{j}$ evaluate $\int (\nabla \wedge \underline{A}).\underline{n}dS$ over the triangle in the x,y-plane bounded by the line $y = 0$, $x = a$ and $y = x$.

Use Stokes's theorem to find the value of the line integral $\int \underline{A}.d\underline{r}$ taken from $x = 0$ to $x = a$ along the hypotenuse of the triangle described in the first part.

17. Evaluate the line integral $I = \oint_C \underline{F}.d\underline{r}$, where $\underline{F} = (x^2+y^2, xy, 0)$ and C is the closed plane curve composed of the arcs

$$x^2 + y^2 = 1 \qquad x \geqslant 0,\ y \geqslant 0,$$
$$x = 0 \qquad (-2 \leqslant y \leqslant 1),$$
$$y = 2(x - 1) \qquad (0 \leqslant x \leqslant 1)\ .$$

Check your answer by converting I into a surface integral.

18. Verify Stokes's theorem for $\underline{A} = (y - z + 2)\underline{i} + (yz+4)\underline{j} - xz\underline{k})$, when S is the surface of the cube $x = 0$, $y = 0$, $z = 0$, $x = 2$, $y = 2$, $z = 2$ above the x,y-plane.

19. Evaluate $\int_S (\nabla \wedge \underline{A}).\underline{n}dS$, where $\underline{A} = (x^2 + y - 4)\underline{i} + 3xy\,\underline{j} + (2x + z^2)\underline{k}$ and S is the surface of:

a) The hemisphere $x^2 + y^2 + z^2 = 16$ above the x,y-plane.

b) The paraboloid $z = 4 - (x^2 + y^2)$ above the x,y-plane.

20. The value at any point P of a vector function $\underline{A}$ is given as $\oint_C r^2\, d\underline{r}$, where r is the distance from P of the element $\delta\underline{r}$ to a fixed closed curve. Show that curl $\underline{A}$ is constant and equal to $-4\int \underline{n}\, dS$, where S is any surface bounded by C.

CHAPTER 3: GENERAL ORTHOGONAL CURVILINEAR COORDINATES

1. Coordinates of position in common systems

1.1. Cartesian system of coordinates

1. In cartesian coordinates, the position of the point is defined by the intersection of three perpendicular planes; x = constant, y = constant, z = constant. Sometimes it is preferable to use other types of coordinates in which the position of the point is defined by the intersection of three surfaces that are not necessarily plane. Problems involving axial symmetry, for instance, are usually dealt with using cylindrical coordinates, while problems with symmetry about a point are best investigated with spherical polar coordinates.

1.2. Cylindrical system of coordinates

In cylindrical coórdinates, the point $A(\rho, \phi, z)$ is defined by the intersection of the cylinder ρ = constant, and the two planes ϕ = constant, z = constant. The unit vectors $\hat{\underline{\rho}}$, $\hat{\underline{\phi}}$, $\hat{\underline{z}} = \underline{k}$ are perpendicular to these surfaces in the directions ρ, ϕ, z increasing respectively. The element of volume bounded by these surfaces at $A(\rho, \phi, z)$ and the neighbouring

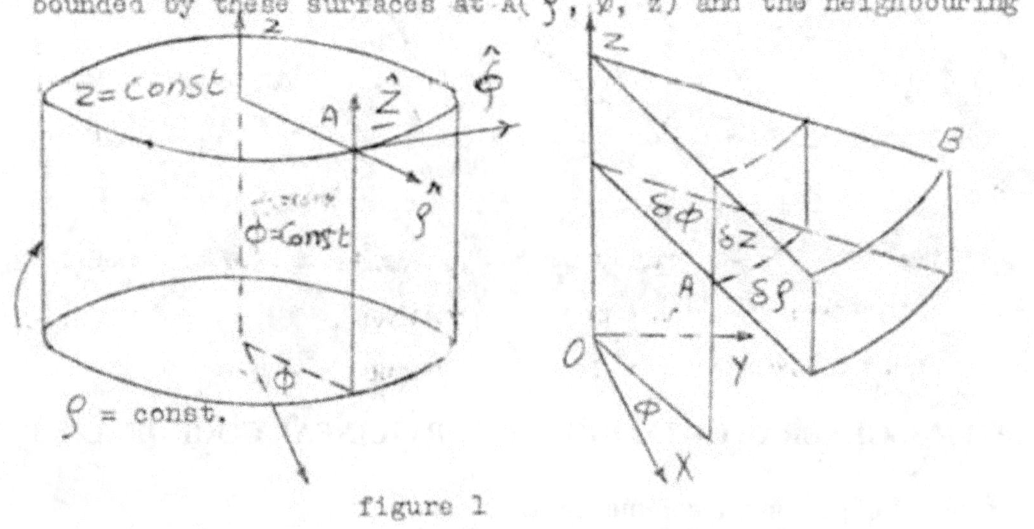

figure 1

point $B(\rho + \delta\rho, \phi + \delta\phi, z + \delta z)$ is shown in figure 1.

That is in the direction of the unit vectors i_1, i_2, i_3.
then

$$\bar{F} = f_1 i_1 + f_2 i_2 + f_3 i_3$$

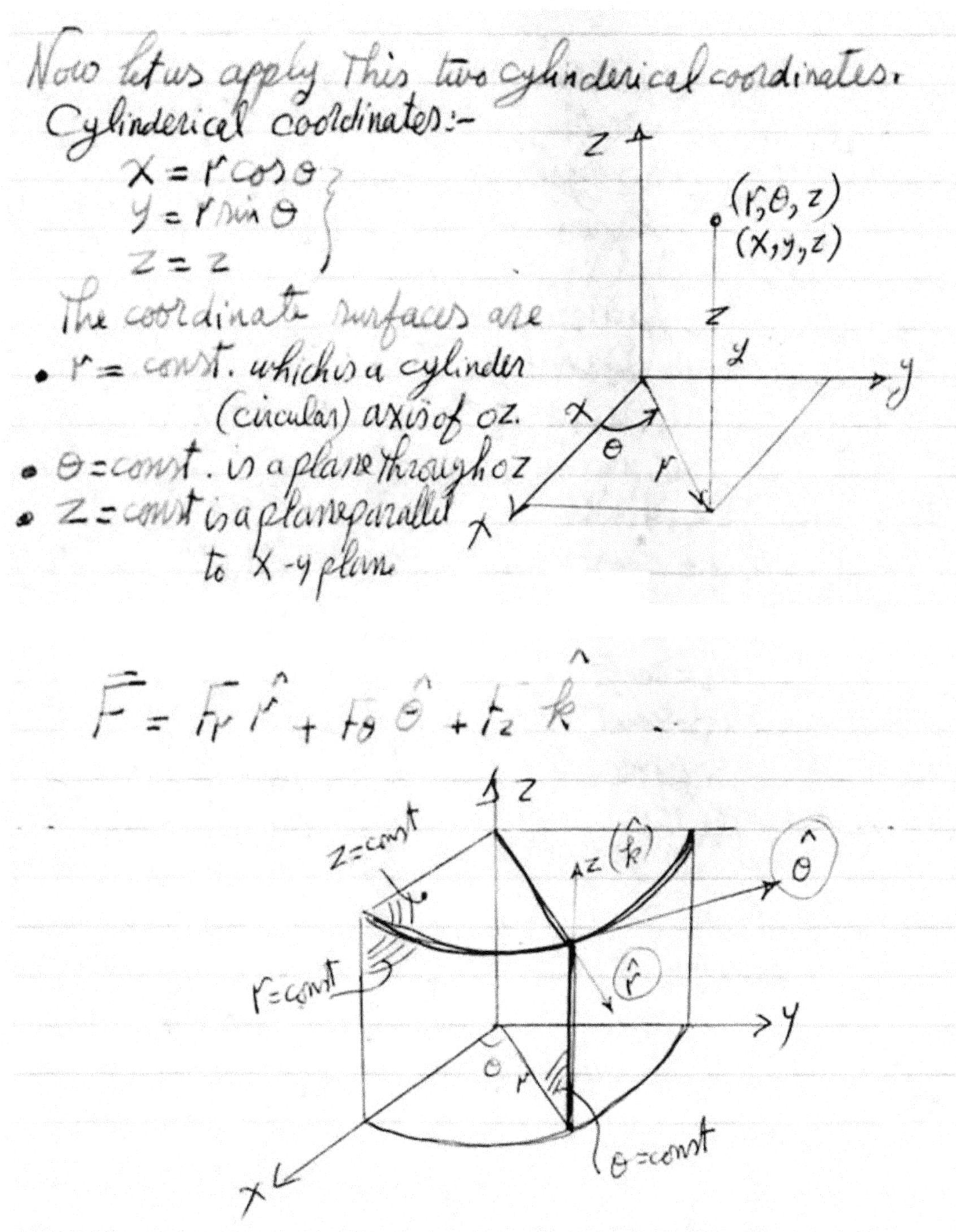

Now let us apply this two cylinderical coordinates.

Cylinderical coordinates:-

$$\left.\begin{aligned} x &= r\cos\theta \\ y &= r\sin\theta \\ z &= z \end{aligned}\right\}$$

The coordinate surfaces are

- $r =$ const. which is a cylinder (circular) axis of oz.
- $\theta =$ const. is a plane through oz
- $z =$ const is a plane parallel to x-y plane

$$\bar{F} = F_r \hat{r} + F_\theta \hat{\theta} + F_z \hat{k}$$

1.3. Spherical system of coordinates

In the spherical polar coordinates, the point A(, Θ, Ø) is defined by the intersection of the sphere r = constant, the cone Θ = constant, and the plane Ø = constant. The unit vectors $\hat{\underline{r}}$, Θ, $\hat{\underline{Ø}}$ are normal to these surfaces in the direction r, Θ, Ø increasing, respectively. The element of volume bounded by the coordinate surfaces at A(r, Θ, Ø) and the neighbouring point B(r + δ r, Θ + δ Θ, Ø + δ Ø) is shown in figure 2.

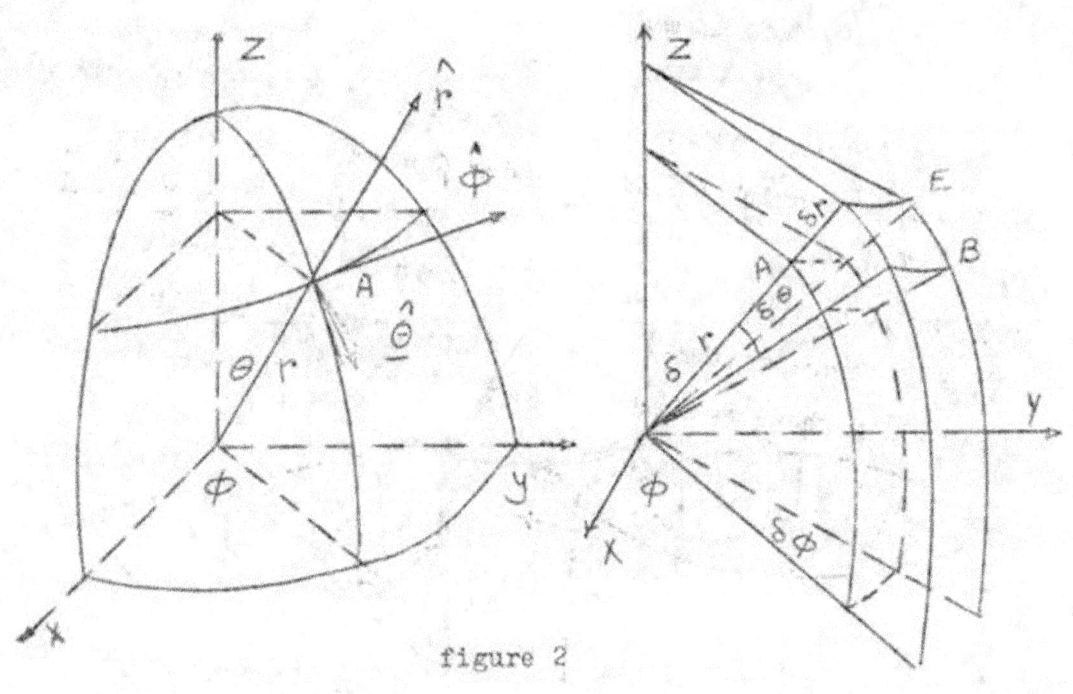

figure 2

Spherical polar coordinates:-

$$x = r \sin\theta \cos\phi$$
$$y = r \sin\theta \sin\theta$$
$$z = r \cos\theta$$

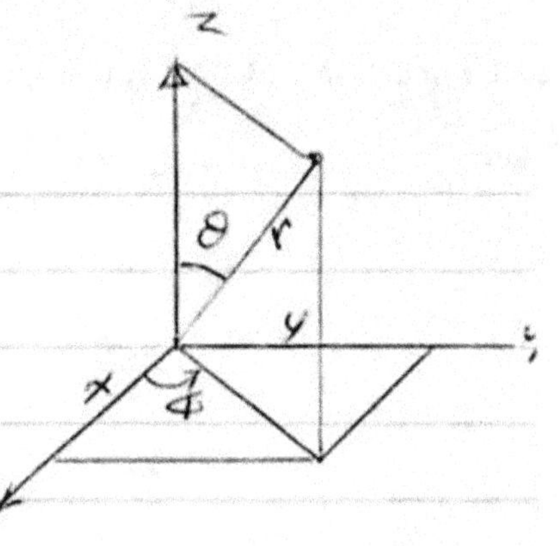

The coordinates surfaces are:-

- r = const. which is a sphere centre origin
- θ = const. which is a cone axis (OZ)
- ϕ = const. which is a plane through (OZ)

$$\vec{F} = F_r \hat{r} + F_\theta \hat{\theta} + F_\phi \hat{\phi}$$

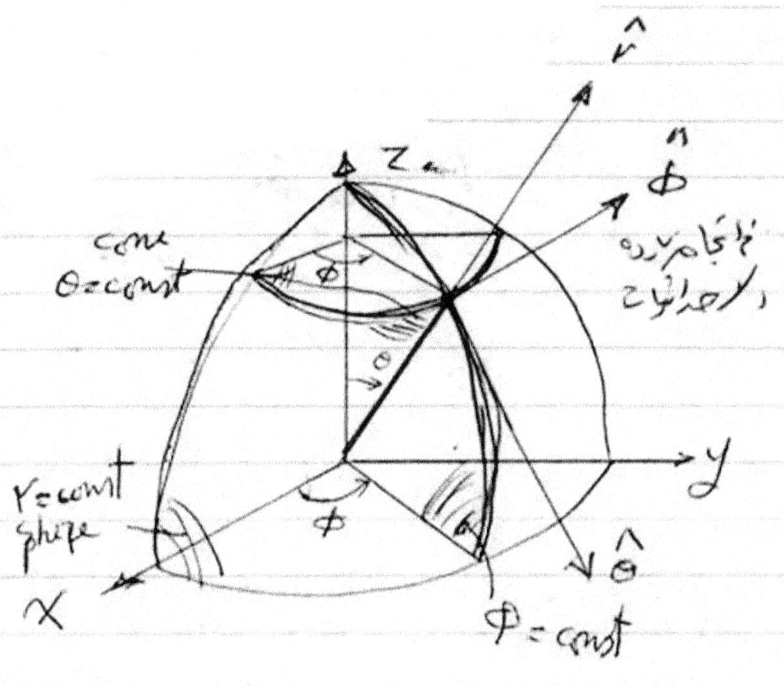

2. Coordinates of position in general systems

2. In general a point P is space can be defined by three surfaces u = constant, v = constant, w = constant where u,v,w are single valued continuous function of x,y,z.

These surfaces are assumed not to coincide or intersect in a common curve. The values of u, v, w on these surfaces may be considered as the coordinates of the point P. The surfaces themselves are called the coordinate surfaces through P, and the curves of intersection the coordinate lines, the tangents

at P of the coordinate lines are the coordinate axes. In general these axes vary from one point to another. When these axes are mutually orthogonal u, v, w are said to constitute a system of orthogonal curvilinear coordinates.

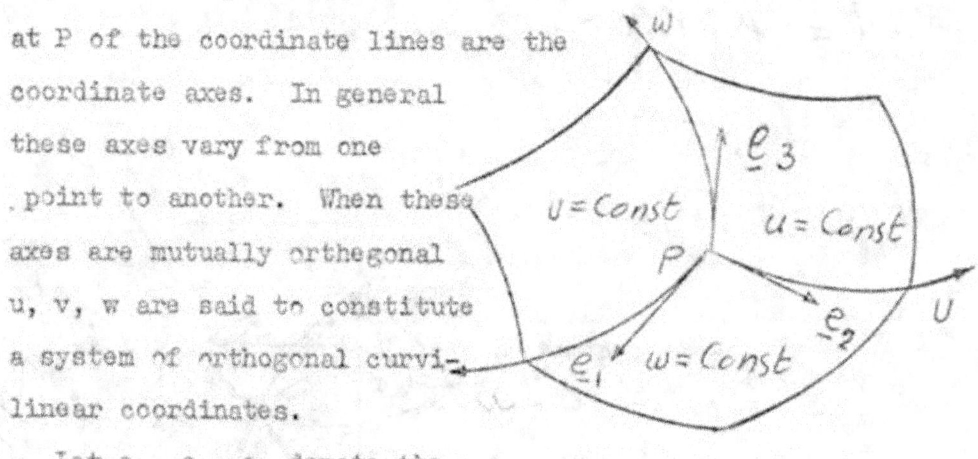

Let $\underline{e}_1$, $\underline{e}_2$, $\underline{e}_3$ denote the

unit vectors normal to the
surfaces u = constant,
v = constant, w = constant
in the direction u, v, w increasing; these are not constant vectors. For an orthogonal system of co-ordinates, these unit vectors are along the coordinate axes. We will choose u,v,w such that $\underline{e}_1$, $\underline{e}_2$, $\underline{e}_3$ will form a right-handed system.

20\1\1972

Orthogonal curvilinear coordinates

In all our radius work the various expressions where obtained in rectangular cartesian coordinates we now consider a new system of coordinates of much wide generality known as the Orthogonal curvilinear coordinates.

Let u_1, u_2, u_3 be a new system of coordinates refered to the curvilinear coordindtes x, y, z by the relations

$$x = f_1(u_1, u_2, u_3)$$
$$y = f_2(u_1, u_2, u_3)$$
$$z = f_2(u_1, u_2, u_3)$$

The surfaces $u_1 =$ const are called coordinate surfaces
$u_2 =$ const
$u_3 =$ const

we shall consider the case when these surfaces are mutually orthogonal. The three surfaces intersect in pairs in three curves which called coordinate lines

Let i_1, i_2, i_3 be the unit vectors along the tangents of the three curves (coordinate lines) now let F be a vector of the field at P & let F_1, F_2, F_3 be the scalar components of vector F along the tangents at P to the coordinate lines

3. The differential volume in general

3. Let u, v, w constitute an orthogonal system of coordinates. Let $\delta\tau$ be the element of volume bounded by the coordinate surfaces, at P(u,v,w) and the neighbouring point P'(u+ δu, v + δ v, w+δ w). On PA, u varies from u at P to u + δ u

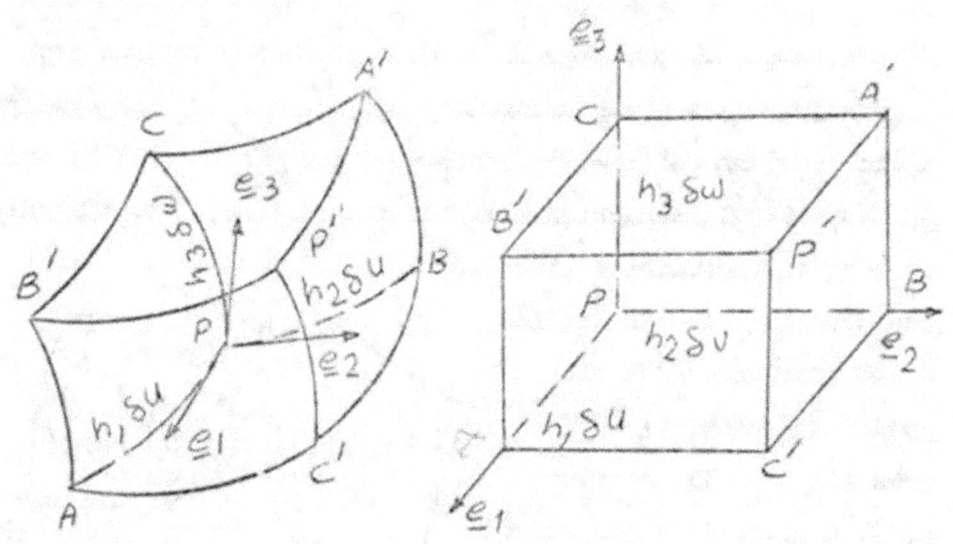

at A. The length of the arc can be put in the form $h_1 \delta u$. Similarly the arcs PB, PC can be put in the forms $h_2 \delta v$, $h_3 \delta w$, respectively. h_1, h_2, h_3 are functions of u, v, w. Consequently

$$\text{the arc } CB' = h_1(u, v, w + \delta w)\delta u = h_1(u, v, w)\delta w$$

$$\frac{\partial h_1}{\partial w}(u,v,w)\delta w\delta u + \ldots$$

$$\text{The arc } BC' = h_1(u,v,+\delta v,w)\delta u = h_1(u,v,w) + \frac{v\,\partial h_1}{\partial v}(u,v,w)\delta v\delta u + \ldots$$

$$\text{The arc } A'P' = h_1(u,v + \delta v, w + \delta w)\delta u$$

$$= h_1(u,v,w)\delta u + \frac{\partial h_1}{\partial v}(u,v,w)\delta v\delta u + \frac{\partial h_1}{\partial w}(u,v,w)\delta w\delta u + \ldots$$

Therefore, the arcs PA, CB', BC', A'P' and similarly the arcs PB, AC', CA', B'P' and the arcs PC, AB', B'A, C'P' are equal to the first order of small quantities. When δu, δv, δw tend to zero, the curvilinear figure PP' approximates to a rectangular parallelepiped.

The volume of the element $\delta\tau = h_1h_2h_3\,\delta u\delta v\delta w$ to the third order of small quantities.

The arc length $PP' = \delta s = \sqrt{(h_1\delta u)^2+(h_2\delta v)^2+(h_3\delta w)^2}$

and the vector $PP' = \underline{\delta s} = \underline{e}_1h_1\delta u + \underline{e}_2h_2\delta v + \underline{e}_3h_3\delta w$, both to the first order of small quantities.

The Differential volume element:-

Suppose that as a result of increasing u_1, u_2, u_3 by du_1, du_2, du_3 that a volume element is formed of dimensions $h_1 du_1$, $h_2 du_2$, $h_3 du_3$.

$\overline{dl} = h_1 du_1 \bar{i}_1 + h_2 du_2 \bar{i}_2 + h_3 du_3 \bar{i}_3$

$dl^2 = h_1^2 du_1^2 + h_2^2 du_2^2 + h_3^2 du_3^2$

volume of element =
$h_1 h_2 h_3 du_1 du_2 du_3$

Let us apply this on cylindrical coordinations.

3.1. Cartesian differential volume

4. In cartesian coordinates take x for u, y for v, z for w. Then $h_1 = h_2 = h_3 = 1$ and $(\delta s)^2 = (\delta x)^2 + (\delta y)^2 + (\delta z)^2$, and the element of volume in these coordinates $\delta \tau = \delta x \delta y \delta z$.

3.2. Cylindrical differential volume

In cylindrical coordinates take ρ for u, ϕ for v, z for w. From figure 1, the element of volume has edges of lengths $\delta\rho$, $\rho\delta\phi$, δz. Therefore, $h_1 = 1$, $h_2 = \rho$, $h_3 = 1$.

$$\therefore \quad (\delta s)^2 = (\delta \rho)^2 + (\rho \delta \phi)^2 + (\delta z)^2$$

$$\delta \tau = \rho \delta \rho \delta \phi \delta z.$$

Cylinderical coordinates

The dimensions of the elements are $(dr, r d\theta, dz)$ hence for cylinderical coordinates.

$u_1 = r$
$u_2 = \theta$
$u_3 = z$
$h_1 = 1$
$h_2 = r$
$h_3 = 1$

3.3. Spherical differential volume

In spherical coordinates take r for u, θ for v, Ø for w. From figure 2, the element of volume at P has edges of lengths δr, $r\delta\theta$, $r \sin\theta\delta\emptyset$, from which we can deduce that $h_1 = 1$, $h_2 = r$, $h_3 = r \sin\theta$.

$$\therefore \quad (\delta s)^2 = (\delta r)^2 + (r\delta\theta)^2 + (r \sin\theta\delta\emptyset)^2,$$

$$\delta\tau = r^2 \sin\theta\,\delta r\delta\theta\delta\emptyset.$$

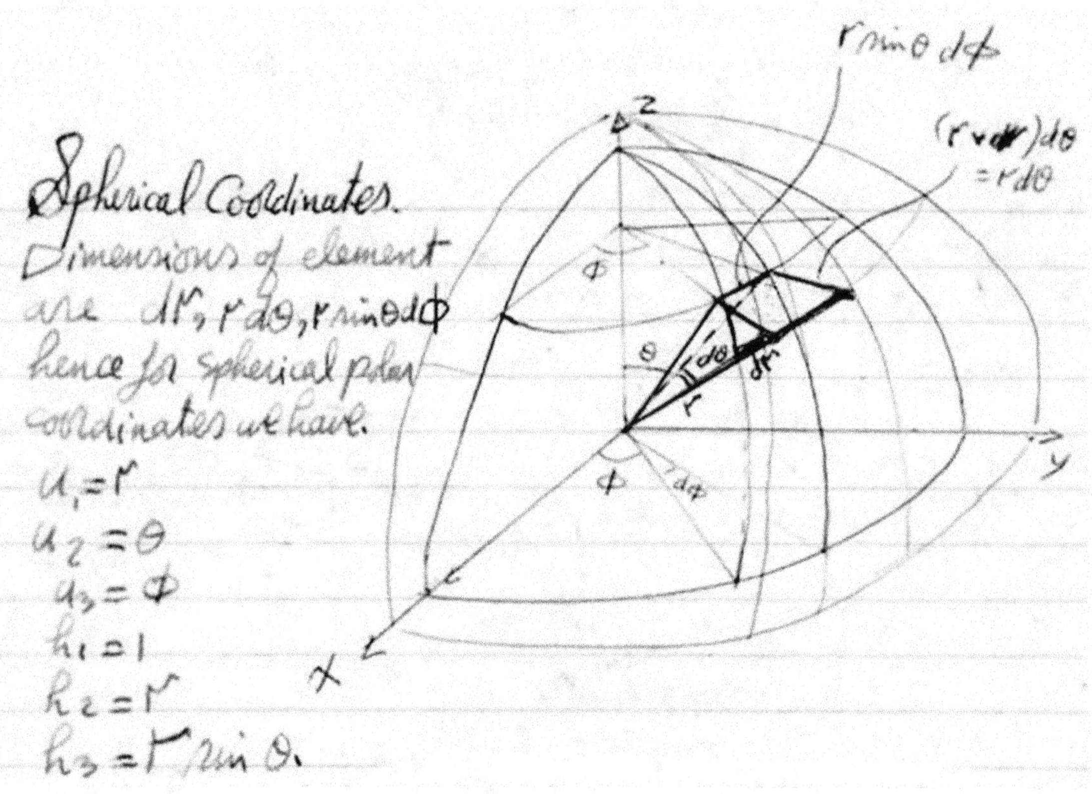

The gradient scalar fn ψ in orthogonal curvilinear coordinates here we use the fundamental property of gradient namely the scalar component of the gradient in any direction is the rate of change of the ψ in that direction

$$\nabla\psi = \frac{1}{h_1}\frac{\partial\psi}{\partial u_1} i_1 + \frac{1}{h_2}\frac{\partial\psi}{\partial u_2} i_2 + \frac{1}{h_3}\frac{\partial\psi}{\partial u_3} i_3$$

3.4. Transformation formulae of differential volumes in different systems of coordinates

5. In the cartesian, cylindrical, and spherical coordinates, it was possible to draw the elements of volume from which h_1, h_2, h_3 were calculated. In other types of coordinates u,v,w as well as these coordinates h_1, h_2, h_3 can be calculated when the transformation formulae relating u,v,w to the cartesian coordinates are known. Let these formulae be

$$\left.\begin{array}{l} x = x(u,v,w),\ y = y(u,v,w),\ z = z(u,v,w) \\ \text{or} \\ \quad \underline{r} = \underline{r}(u,\ v,\ w) \end{array}\right\} (1)$$

On the u coordinate curve v = constant, w = constant. Its parametric equation is, therefore, given by (1) with u as the parameter. The tangent at (u, v, w) has direction ratios

$$\frac{\partial x}{\partial u}, \frac{\partial y}{\partial u}, \frac{\partial z}{\partial u} \quad ,$$

or direction cosines

$$\frac{\frac{\partial x}{\partial u}, \frac{\partial y}{\partial u}, \frac{\partial z}{\partial u}}{\sqrt{(\frac{\partial x}{\partial u})^2 + (\frac{\partial y}{\partial u})^2 + (\frac{\partial z}{\partial u})^2}} .$$

Therefore, the unit vector $\underline{e}_1$ is given by

$$\underline{e}_1 = \frac{\underline{i}\frac{\partial x}{\partial u} + \underline{j}\frac{\partial y}{\partial u} + \underline{k}\frac{\partial z}{\partial u}}{\sqrt{(\frac{\partial x}{\partial u})^2 + (\frac{\partial y}{\partial u})^2 + (\frac{\partial z}{\partial u})^2}} = \frac{\frac{\partial \underline{r}}{\partial u}}{\left|\frac{\partial \underline{r}}{\partial u}\right|}$$

Similarly

$$\underline{e}_2 = \frac{\frac{\partial \underline{r}}{\partial v}}{\left|\frac{\partial \underline{r}}{\partial v}\right|} \quad , \quad \underline{e}_3 = \frac{\frac{\partial \underline{r}}{\partial w}}{\left|\frac{\partial \underline{r}}{\partial w}\right|}$$

If $\underline{e}_1$, $\underline{e}_2$, $\underline{e}_3$ are mutually orthogonal, then

$$\underline{e}_1 \cdot \underline{e}_2 = 0, \underline{e}_2 \cdot \underline{e}_3 = 0, \underline{e}_3 \cdot \underline{e}_1 = 0$$

i.e. $\frac{\partial \underline{r}}{\partial u} \cdot \frac{\partial \underline{r}}{\partial v} = 0, \frac{\partial \underline{r}}{\partial v} \cdot \frac{\partial \underline{r}}{\partial w} = 0, \frac{\partial \underline{r}}{\partial w} \cdot \frac{\partial \underline{r}}{\partial u} = 0 .$

The element of arc δs is given by:

$$(\delta s)^2 = (\delta x)^2 + (\delta y)^2 + (\delta z)^2 = (\delta \underline{r})^2$$

$$= (\frac{\partial \underline{r}}{\partial u}\delta u + \frac{\partial \underline{r}}{\partial v}\delta v + \frac{\partial \underline{r}}{\partial w}\delta w)^2$$

$$= (\frac{\partial \underline{r}}{\partial u})^2(\delta u)^2 + (\frac{\partial \underline{r}}{\partial v})^2(\delta v)^2 + (\frac{\partial \underline{r}}{\partial w})^2(\delta w)^2$$

$$+2\frac{\partial \underline{r}}{\partial u} \cdot \frac{\partial \underline{r}}{\partial v}\delta u \delta v + 2\frac{\partial \underline{r}}{\partial v} \cdot \frac{\partial \underline{r}}{\partial w}\delta r \delta w$$

$$+ 2\frac{\partial \underline{r}}{\partial u}\delta w \delta u.$$

If the coordinates are mutually orthogonal, then

$$(\delta s)^2 = h_1^2(\delta u)^2 + h_2^2(\delta v)^2 + h_3^2(\delta w)^2$$

where

$$h_1^2 = \left(\frac{\partial \underline{r}}{\partial u}\right)^2 = \left(\frac{\partial x}{\partial u}\right)^2 + \left(\frac{\partial y}{\partial u}\right)^2 + \left(\frac{\partial z}{\partial u}\right)^2,$$

$$h_2^2 = \left(\frac{\partial \underline{r}}{\partial v}\right)^2 = \left(\frac{\partial x}{\partial v}\right)^2 + \left(\frac{\partial y}{\partial v}\right)^2 + \left(\frac{\partial z}{\partial v}\right)^2,$$

$$h_3^2 = \left(\frac{\partial \underline{r}}{\partial w}\right)^2 = \left(\frac{\partial x}{\partial w}\right)^2 + \left(\frac{\partial y}{\partial w}\right)^2 + \left(\frac{\partial z}{\partial w}\right)^2.$$

4. grad, div, and cur

6. Grad ϕ, div $\underline{v}$, curl $\underline{v}$ and $\nabla^2\phi$ in orthogonal curvilinear coordinates.

i) Grad ϕ.

The scalar component of $\nabla\phi$ in the direction $\hat{\underline{s}} = \hat{\underline{s}}\cdot\nabla\phi$

$$= \frac{\partial \phi}{\partial s} = \lim_{\delta s \to 0} \frac{\delta\phi}{\delta s}$$

where $\delta\phi$ is the change in ϕ associated with δs.

$\therefore$ Components of $\nabla\phi$ in the directions $\underline{e}_1, \underline{e}_2, \underline{e}_3$ are respectively

$$\lim_{\delta u \to 0} \frac{\delta\phi}{h_1\delta u}, \quad \lim_{\delta v \to 0} \frac{\delta\phi}{h_2\,\delta v}, \quad \lim_{\delta w \to 0} \frac{\delta\phi}{h_3\delta w},$$

i.e. $\dfrac{1}{h_1}\dfrac{\partial\phi}{\partial u}, \dfrac{1}{h_2}\dfrac{\partial\phi}{\partial v}, \dfrac{1}{h_3}\dfrac{\partial\phi}{\partial w}$.

4.1. Gradient

Therefore,

$$\text{grad } \phi = \frac{\underline{e}_1}{h_1}\frac{\partial \phi}{\partial u} + \frac{\underline{e}_2}{h_2}\frac{\partial \phi}{\partial v} + \frac{\underline{e}_3}{h_3}\frac{\partial \phi}{\partial w} .$$

4.2. Divergence

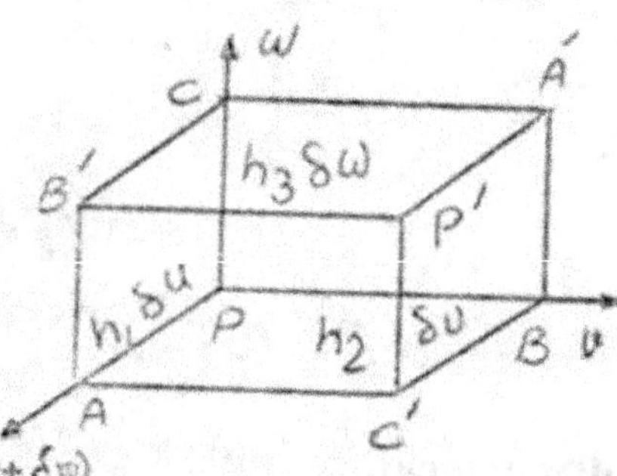

ii) Div $\underline{V}$

Apply Gauss's theorem to the element of volume bounded by the coordinate planes at $P(u,v,w)$ and $P'(u+\delta u, v+\delta v, w+\delta w)$

$$\therefore \int_{\text{element}} \nabla \cdot \underline{V}\, d = \left\{(\nabla \cdot \underline{V}) h_1 h_2 h_3\right\}_P \delta u \delta v \delta w + \ldots$$

$$\int_{PAB'C} \underline{V}.\underline{n}\, dS = -(V_2 h_1 h_3)_P \delta u \delta w + \ldots.$$

$$\int_{BC'P'A'} \underline{V}.\underline{n}\, dS = (V_2 h_1 h_3)_B \delta u \delta w + \ldots.$$

$$=(V_2 h_1 h_3)_P \delta u \delta w + \left\{\frac{\partial}{\partial v}(V_2 h_1 h_3)\right\}_P \delta u \delta v \delta w + \ldots$$

$$\therefore \int_{PAB'C} \underline{V}.\underline{n}dS + \int_{BC'P'A'} \underline{V}.\underline{n}dS = \left\{\frac{\partial}{\partial v}(V_2 h_1 h_3)\right\}_P \delta u \delta v \delta w + \ldots$$

Similarly

$$\int_{PBA'C} \underline{V}.\underline{n}dS + \int_{AC'P'B'} \underline{V}.\underline{n}dS = \left\{\frac{\partial}{\partial u}(V_1 h_2 h_3)\right\}_P \delta u \delta v \delta w + \ldots$$

and
$$\int_{PAC'B} \underline{V}.\underline{n}dS + \int_{CB'P'A'} \underline{V}.\underline{n}dS = \left\{\frac{\partial}{\partial w}(V_1 h_2 h_3)\right\}_P \delta u \delta v \delta w + \ldots$$

Gauss's theorem gives

$$\left\{(\nabla.\underline{V})h_1h_2h_3\right\}_P \delta u\delta v\delta w = \left\{\frac{\partial}{\partial u}(h_2h_3V_1) + \frac{\partial}{\partial v}(h_3h_1V_2) + \frac{\partial}{\partial w}(h_1h_2V_3)\right\}_P \delta u\delta v\delta w + \cdots$$

Divide by $\delta u\delta v\delta w$ and

let the element shrink to P we get

$$\text{div } \underline{V} = \frac{1}{h_1h_2h_3}\left[\frac{\partial}{\partial u}(h_2h_3V_1) + \frac{\partial}{\partial v}(h_3h_1V_2) + \frac{\partial}{\partial w}(h_1h_2V_3)\right]$$

12\2\472

The divergence of a vector in Orthogonal curvilinear coordinates.

Here ~~After~~ we use the property of the divergence namely that the divergence at a point is equal to the flux diverging from the element per unit volum when the volume shrinks to a point flux intering the element $= F_1 h_2 h_3\, du_2 du_3$

" diverging from the ele $= F_1 h_2 h_3\, du_2 du_3 + \frac{\partial}{\partial u_1}(F_1 h_2 h_3\, du_2 du_3)\, du_1$

surface ②

Netflux diverging $= \frac{\partial}{\partial u_1}(F_1 h_2 h_3)\, du_1\, du_2\, du_3$

adding two similar expressions and dividing by the volume of the element.

$$\nabla \cdot F = \frac{1}{h_1 h_2 h_3}\left[\frac{\partial}{\partial u_1}(F_1 h_2 h_3) + \frac{\partial}{\partial u_2}(F_2 h_3 h_1) + \frac{\partial}{\partial u_3}(F_3 h_1 h_2)\right]$$

$$\nabla \cdot \bar{F} = \frac{1}{h_1 h_2 h_3}\left[\frac{\partial}{\partial u_1}(F_1 h_2 h_3) + \frac{\partial}{\partial u_2}(F_2 h_1 h_3) + \frac{\partial}{\partial u_3}(F_3 h_1 h_2)\right]$$

Cylinderical coordinates

$u_1 = r \qquad h_1 = 1$

$u_2 = \theta \qquad h_2 = r$

$u_3 = z \qquad h_3 = 1$

$$\nabla \cdot \bar{F} = \frac{1}{r}\left[\frac{\partial}{\partial r}(F_r r) + \frac{\partial}{\partial \theta}(F_\theta) + \frac{\partial}{\partial z}(F_z r)\right]$$

$$= \frac{1}{r}\frac{\partial}{\partial r}(r F_r) + \frac{1}{r}\frac{\partial}{\partial \theta}(F_\theta) + \frac{\partial F_z}{\partial z}.$$

Spherical polar coordinates.

$u_1 = r \qquad h_1 = 1$

$u_2 = \theta \qquad h_2 = r$

$u_3 = \phi \qquad h_3 = r \sin\theta$

$$\nabla \cdot \bar{F} = \frac{1}{r^2 \sin\theta}\left[\frac{\partial}{\partial r}(F_r r^2 \sin\theta) + \frac{\partial}{\partial \theta}(F_\theta r \sin\theta) + \frac{\partial}{\partial \phi}(F_\phi)\right]$$

$$\frac{\partial}{h_1 \partial u_1}(F_1 h_2 h_3 \, du_2 \, du_3) h_1 \, du_1 = \frac{\partial}{\partial u_1}(F_1 h_2 h_3 \, du_2 \, du_3$$

The Laplacian operator $\nabla^2\psi$ in the orthogonal curvilinear coordinates:–

$$\nabla^2\psi = \nabla\cdot(\nabla\psi)$$

Now $\nabla\psi$ has scalar components $\frac{1}{h_1}\frac{\partial\psi}{\partial u_1}, \frac{1}{h_2}\frac{\partial\psi}{\partial u_2}, \frac{1}{h_3}\frac{\partial\psi}{\partial u_3}$

$$\therefore \nabla^2\psi = \frac{1}{h_1h_2h_3}\left[\frac{\partial}{\partial u_1}\left(\frac{h_2h_3}{h_1}\frac{\partial\psi}{\partial u_1}\right) + \frac{\partial}{\partial u_2}\left(\frac{h_3h_1}{h_2}\frac{\partial\psi}{\partial u_2}\right) + \frac{\partial}{\partial u_3}\left(\frac{h_1h_2}{h_3}\frac{\partial\psi}{\partial u_3}\right)\right]$$

$$\nabla^2\psi = \frac{1}{h_1h_2h_3}\left[\frac{\partial}{\partial u_1}\left(\frac{h_2h_3}{h_1}\frac{\partial\psi}{\partial u_1}\right) + \frac{\partial}{\partial u_2}\left(\frac{h_1h_3}{h_2}\frac{\partial\psi}{\partial u_2}\right) + \frac{\partial}{\partial u_3}\left(\frac{h_1h_2}{h_3}\frac{\partial\psi}{\partial u_3}\right)\right]$$

Cylinderical coordinates:–

$u_1 = r \qquad h_1 = 1$
$u_2 = \theta \qquad h_2 = r$
$u_3 = Z \qquad h_3 = 1$

$$\nabla^2\psi = \nabla\cdot(\nabla\psi) = \frac{1}{r}\left[\frac{\partial}{\partial r}\left(r\frac{\partial\psi}{\partial r}\right) + \frac{\partial}{\partial\theta}\left(\frac{1}{r}\frac{\partial\psi}{\partial\theta}\right) + \frac{\partial}{\partial z}\left(r\frac{\partial\psi}{\partial z}\right)\right]$$

$$= \frac{\partial^2\psi}{\partial r^2} + \frac{1}{r}\frac{\partial\psi}{\partial r} + \frac{1}{r^2}\frac{\partial^2\psi}{\partial\theta^2} + \frac{\partial^2\psi}{\partial z^2}.$$

Spherical polar coordinates:-

$u_1 = r \qquad h_1 = 1$

$u_2 = \theta \qquad h_2 = r$

$u_3 = \phi \qquad h_3 = r\sin\theta$

$$\nabla^2\psi = \nabla\cdot(\nabla\psi)$$

$$= \frac{1}{r^2\sin\theta}\left[\frac{\partial}{\partial r}\left(r^2\sin\theta\frac{\partial\psi}{\partial r}\right)+\frac{\partial}{\partial\theta}\left(\sin\theta\frac{\partial\psi}{\partial\theta}\right)+\frac{\partial}{\partial\phi}\left(\frac{1}{\sin\theta}\frac{\partial\psi}{\partial\phi}\right)\right]$$

$$= \frac{1}{r^2}\frac{\partial}{\partial r}\left(r^2\frac{\partial\psi}{\partial r}\right)+\frac{1}{r^2\sin\theta}\frac{\partial}{\partial\theta}\left(\sin\theta\frac{\partial\psi}{\partial\theta}\right)+\frac{1}{r^2\sin^2\theta}\frac{\partial^2\psi}{\partial\phi^2}$$

Cylinderical coordinates:- $h_1 = 1 \quad h_2 = r \quad h_3 = 1$

$$\nabla\psi = \frac{\partial\psi}{\partial r}\hat{r} + \frac{1}{r}\frac{\partial\psi}{\partial\theta}\hat{\theta} + \frac{\partial\psi}{\partial z}\hat{k}$$

Spherical polar coordinates

$h_r = 1 \qquad h_\theta = r \qquad h_\phi = r\sin\theta$

$$\nabla\psi = \frac{\partial\psi}{\partial r}\hat{r} + \frac{1}{r}\frac{\partial\psi}{\partial\theta}\hat{\theta} + \frac{1}{r\sin\theta}\frac{\partial\psi}{\partial\phi}\hat{\phi}$$

4.3. Curl or rotation

iii) Curl $\underline{V}$

Apply Stokes's theorem to the side PBA'C of the element

$$\int_{PBA'C} (\nabla\wedge\underline{V}).\underline{n}\, dS = \left\{(\nabla\wedge V)h_2h_3\right\}_P \delta v\, \delta w + \ldots$$

where $(\nabla\wedge\underline{V})_1$ is the component of $\nabla\wedge\underline{V}$ in the direction $\underline{e}_1$.

w, C, A', $h_3\delta w$, $h_2\delta v$, P, B, v, u

$$\int_{PB} \underline{V}.d\underline{r} = (v_2h_2)_P\delta v + \ldots$$

$$\int_{CA'} \underline{V}.d\underline{r} = (v_2h_2)_C\delta v + \ldots = (v_2h_2)_P\delta v + \left.\frac{\partial}{\partial w}(v_2h_2)\right\}_P \delta v\delta w + \ldots$$

$$\int_{PC} \underline{V}.d\underline{r} = (v_3h_3)_P\delta w + \ldots$$

$$\int_{BA'} \underline{V}.d\underline{r} = (v_3h_3)_B\delta w + \ldots = (v_3h_3)_P\delta w + \left\{\frac{\partial}{\partial v}(v_3h_3)\right\}_P \delta v\delta w + \ldots$$

$$\therefore \int_{PBA'C} \underline{V}.d\underline{r} = \int_{PB} \underline{V}.d\underline{r} + \int_{CA'} \underline{V}.d\underline{r} + \int_{BA'} \underline{V}.d\underline{r} + \int_{CP} \underline{V}.d\underline{r}$$

$$= \left\{\frac{\partial}{\partial v}(V_3h_3) - \frac{\partial}{\partial w}(V_2h_2)\right\}_P \delta v\delta w + \ldots$$

Stoke's theorem gives

$$\left\{(\nabla\wedge\underline{V})_1\, h_2h_3\right\}_P \delta v\,\delta w = \left\{\frac{\partial}{\partial v}(V_3h_3) - \frac{\partial}{\partial w}(V_2h_2)\right\}_P \delta v\,\delta w + \ldots$$

Divide by $\delta v\,\delta w$ and let the element shrink to P, we get

$$(\nabla\wedge\underline{V})_1 = \frac{1}{h_2h_3}\left\{-(\frac{\partial}{\partial w}(V_2h_2) + \frac{\partial}{\partial v}(V_3h_3)\right\}$$

Similarly

$$(\nabla\wedge\underline{V})_3 = \frac{1}{h_1h_2}\left\{\frac{\partial}{\partial u}(V_2h_2) - \frac{\partial}{\partial v}(V_1h_1)\right\} \text{ and}$$

$$(\nabla\wedge\underline{V})_2 = \frac{1}{h_3h_1}\left\{\frac{\partial}{\partial w}(V_1h_1) - \frac{\partial}{\partial u}(V_3h_3)\right\}.$$

Therefore

$$\nabla \wedge \underline{V} = \frac{1}{h_1 h_2 h_3} \Big[\underline{e}_1 h_1 \Big\{ \frac{\partial}{\partial v} (V_3 h_3) - \frac{\partial}{\partial w} (V_2 h_2) \Big\}$$

$$+ \underline{e}_2 h_2 \Big\{ \frac{\partial}{\partial w} (V_1 h_1) - \frac{\partial}{\partial u} (V_3 h_3) \Big\} + \underline{e}_3 h_3 \Big\{ \frac{\partial}{\partial u} (V_2 h_2)$$

$$- \frac{\partial}{\partial v} (V_1 h_1) \Big\} \Big].$$

$$\text{i.e. Curl } \underline{V} = \frac{1}{h_1 h_2 h_3} \begin{vmatrix} h_1 \underline{e}_1 & h_2 \underline{e}_2 & h_3 \underline{e}_3 \\ \frac{\partial}{\partial u} & \frac{\partial}{\partial v} & \frac{\partial}{\partial w} \\ h_1 v_1 & h_2 v_2 & h_3 v_3 \end{vmatrix} .$$

4.4. Divergence of gradient

iv) $\nabla^2 \phi = \text{div grad } \phi$

$$= \frac{1}{h_1 h_2 h_3} \Big[\frac{\partial}{\partial u} \Big(\frac{h_2 h_3}{h_1} \frac{\partial \phi}{\partial u} \Big)$$

$$+ \frac{\partial}{\partial v} \Big(\frac{h_3 h_1}{h_2} \frac{\partial \phi}{\partial v} \Big)$$

$$+ \frac{\partial}{\partial w} \Big(\frac{h_1 h_2}{h_3} \frac{\partial \phi}{\partial w} \Big) \Big]$$

5. Curl and divergence in general orthogonal systems of coordinates

7. Div $\underline{V}$ and cur $\underline{V}$ can also be obtained by purely algebraical calculations as follows:

From the expression for the gradient we have:

$$\nabla u = \frac{\underline{e}_1}{h_1} \quad , \quad \nabla v = \frac{\underline{e}_2}{h_2} \quad , \quad \nabla w = \frac{\underline{e}_3}{h_3} \; .$$

Then

$$\nabla v \wedge \nabla w = \frac{\underline{e}_2 \wedge \underline{e}_3}{h_2 \, h_3} = \frac{\underline{e}_1}{h_2 h_3} \; .$$

Therefore

$$\underline{e}_1 = h_1 \nabla u = h_2 h_3 \nabla v \wedge \nabla w \; .$$

Similarly

$$\underline{e}_2 = h_2 \nabla v = h_3 h_1 \; \nabla w \wedge \nabla u \; ,$$

$$\underline{e}_3 = h_3 \nabla w = h_1 h_2 \quad u \wedge \nabla v \; .$$

Let $\underline{V} = V_1 \, \underline{e}_1 + V_2 \, \underline{e}_2 + V_3 \, \underline{e}_3$.

Then $\nabla . \underline{V} = \nabla . \, V_1 \underline{e}_1 + \nabla . V_2 \underline{e}_2 + \nabla . V_3 \underline{e}_3$

But $\nabla . V_1 \underline{e}_1 = \nabla . (V_1 h_2 h_3 \nabla v \wedge \nabla w)$

$$= \nabla (V_1 h_2 h_3) . (\nabla v \wedge \nabla w) + v_1 h_2 h_3 \nabla . (\nabla v \wedge \nabla w)$$

$$= \nabla (V_1 h_2 h_3) \; . \; \frac{\underline{e}_1}{h_2 h_3}$$

$$= \left[\frac{\underline{e}_1}{h_1} \frac{\partial}{\partial u} (V_1 h_2 h_3) + \frac{\underline{e}_2}{h_2} \frac{\partial}{\partial v} (V_1 h_3 h_2) + \frac{\underline{e}_3}{h_3} \frac{\partial}{\partial w} (V_1 h_2 h_3) \right] \cdot \frac{\underline{e}_1}{h_2 h_3}$$

$$= \frac{1}{h_1 h_2 h_3} \frac{\partial}{\partial u} (h_2 h_3 V_1) .$$

Similarly

$$\nabla . V_2 \, \underline{e}_2 = \frac{1}{h_1 h_2 h_3} \frac{\partial}{\partial v} (h_3 h_1 V_2) ,$$

$$\nabla \cdot V_3 \, \underline{e}_3 = \frac{1}{h_1 h_2 h_3} \frac{\partial}{\partial w} (h_1 h_2 V_3).$$

Therefore

$$\text{div } \underline{V} = \frac{1}{h_1 h_2 h_3} \left[\frac{\partial}{\partial u} (h_2 h_3 V_1) + \frac{\partial}{\partial v} (h_3 h_1 V_2) + \frac{\partial}{\partial w} (h_1 h_2 V_3) \right].$$

For curl $\underline{V}$ we have:

$$\text{Curl } \underline{V} = \nabla \wedge \underline{V} = \nabla \wedge (V_1 \underline{e}_1 + V_2 \underline{e}_2 + V_3 \underline{e}_3) .$$

But

$$\nabla \wedge (V_1 \underline{e}_1) = \nabla \wedge (h_1 V_1 \nabla u)$$

$$= \nabla (h_1 V_1) \wedge \nabla u + h_1 V_1 \nabla \wedge (\nabla u)$$

$$= \nabla (h_1 V_1) \wedge \frac{\underline{e}_1}{h_1}$$

$$= \left[\frac{\underline{e}_1}{h_1}\frac{\partial}{\partial u}(h_1V_1) + \frac{\underline{e}_2}{h_2}\frac{\partial}{\partial v}(h_1V_1) + \frac{\underline{e}_3}{h_3}\frac{\partial}{\partial w}(h_1V_1)\right] \wedge \frac{\underline{e}_1}{h_1} \cdot$$

$$\therefore \nabla \wedge V_1\underline{e}_1 = \frac{\underline{e}_2}{h_3h_1}\frac{\partial}{\partial w}(h_1V_1) - \frac{\underline{e}_3}{h_1h_2}\partial(h_1V_1)/\partial v \, .$$

Similarly

$$\nabla \wedge V_2\underline{e}_2 = \frac{\underline{e}_3}{h_2h_1}\frac{\partial}{\partial u}(h_2V_2) - \frac{\underline{e}_1}{h_3h_2}\frac{\partial}{\partial w}(h_2V_2) \, ,$$

$$\nabla \wedge V_3\underline{e}_3 = \frac{\underline{e}_1}{h_2h_3}\frac{\partial}{\partial v}(h_3V_3) - \frac{\underline{e}_2}{h_3h_1}\partial(h_3V_3)/\partial u \, .$$

Therefore

$$\text{curl } \underline{V} = \frac{\underline{e}_1}{h_2h_3}\left\{\frac{\partial}{\partial v}(h_3V_3) - \frac{\partial}{\partial w}(h_2V_2)\right\} + \frac{\underline{e}_2}{h_3h_1}\left\{\frac{\partial}{\partial w}(h_1V_1) - \frac{\partial}{\partial u}(h_3V_3)\right\} + \frac{\underline{e}_3}{h_1h_2}\left\{\frac{\partial}{\partial u}(h_2V_2) - \frac{\partial}{\partial v}(h_1V_1)\right\}$$

$$= \frac{1}{h_1h_2h_3}\begin{vmatrix} h_1\underline{e}_1 & h_2\underline{e}_2 & h_3\underline{e}_3 \\ \frac{\partial}{\partial u} & \frac{\partial}{\partial v} & \frac{\partial}{\partial w} \\ h_1V_1 & h_2V_2 & h_3V_3 \end{vmatrix} \, .$$

The curl of the vector in orthogonal curvilinear coordinates

Here we use the fundamental property of the curl of the vector namely the line integral of the vector along the boundary of an infinitesimal area divided by the area is equal - in the limit to the scalar component of the curl of the vector normal to the area we shall now show that

$$\nabla \times \bar{F} = \frac{1}{h_1 h_2 h_3} \begin{vmatrix} h_1 i_1 & h_2 i_2 & h_3 i_3 \\ \frac{\partial}{\partial u_1} & \frac{\partial}{\partial u_2} & \frac{\partial}{\partial u_3} \\ h_1 F_1 & h_2 F_2 & h_3 F_3 \end{vmatrix}$$

Integrate $\bar{F}$ along the boundary

Line integral of $\bar{F}$ along AB, BC, CD, DA

A

B

C

D

$h_3 du_3$

$h_1 du_1$

i_2

$h_2 du_2$

Right hand screw

Line Integral of $\bar{F}$ along $AB = -F_3 h_3 du_3$

~ ~ ~ ~ $CD = F_3 h_3 du_3 + \frac{\partial}{\partial u_2}(F_3 h_3 du_3)\,du_2$

Net $= \frac{\partial}{\partial u_2}(F_3\, h_3\, du_3)\,du_2 \longrightarrow (1)$

Line integral of $\bar{F}$ along $BC = F_2 h_2 du_2$

~ ~ ~ ~ $DC = -F_2 h_2 du_2 - \frac{\partial}{\partial u_3}(F_2 h_2 du_2)\,du_3$

Net $= -\frac{\partial}{\partial u_3}(F_2 h_2\, du_2)\,du_3 \longrightarrow (2)$

total line integral $\bar{F}$ along the Boundary ABCD

$$= \frac{\partial}{\partial u_2}(F_3\, h_3)\,du_2\,du_3 - \frac{\partial}{\partial u_3}(F_2\, h_2)\,du_2\,du_3$$

Divided by the area $h_3\, h_2\, du_2\, du_3$ we get

$$\frac{1}{h_2 h_3}\left[\frac{\partial}{\partial u_2}(F_3 h_3) - \frac{\partial (F_2 h_2)}{\partial u_3}\right]$$

Which is the component of curl F normal to the area abcd that is in the direction of the unit vector $\hat{i}_z$

ex. Given the vector field cylinderical coordinates

$F = r^2 \cos\theta\, \hat{r} + r \sin\theta\, \hat{\theta}$.

i) evaluate the line integral $\oint \bar{F}.\bar{dl}$ over a closed contour which includes the positive y-x-axis & the quadreant of circle of radius

a
ii) Calculate curl $\bar{F}$.

iii) verify Stoke's theory for this problem

Sol^{n}

$$\int_{c} \bar{F}.d\bar{l} = \int_{b} \bar{F}.d\bar{l} + \int_{c} \bar{F}.d\bar{l} + \int_{d} \bar{F}.d\bar{l}.$$

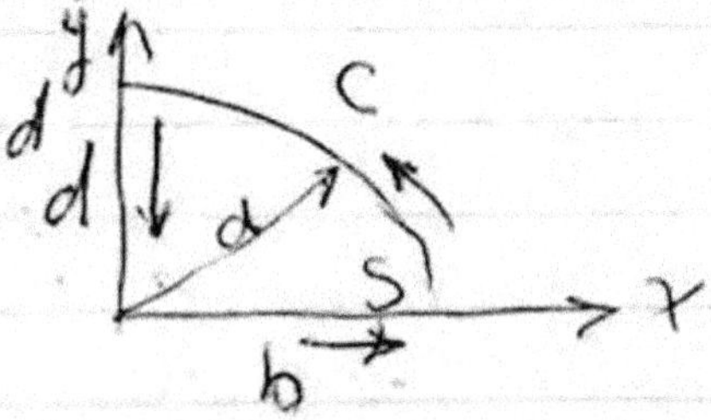

$$\int_{b} \bar{F}.d\bar{l} = \int_{0}^{a} r^2 \cos\theta \, dr \quad \text{with } \theta = 0$$

$$\frac{a^3}{3} \longrightarrow (1)$$

not $rd\theta$

$$\int_{c} \bar{F}.d\bar{l} = \int_{0}^{\frac{\pi}{2}} r \sin\theta \, r d\theta \quad \text{with } r = a$$

$$= a^2 \longrightarrow (2)$$

$$\int_{a} \bar{F}.d\bar{l} = -\int_{r}^{a} r^2 \cos\theta \, dr \quad \text{with } \theta = \frac{\pi}{2}$$

$$= 0 \longrightarrow (3)$$

$$\text{Total} = \int_C F\,d\bar{l} = \frac{a^3}{3} + a^2$$

$$\nabla \times \bar{F} = \frac{1}{r}\begin{vmatrix} \hat{r} & r\hat{\theta} & \hat{k} \\ \frac{\partial}{\partial r} & \frac{\partial}{\partial \theta} & \frac{\partial}{\partial z} \\ r^2\cos\theta & r^2 & 0 \end{vmatrix}$$

$$= \frac{1}{r}\left(2r\sin\theta + r^2\sin\theta\right)\hat{k} = (2+r)\sin\theta\,\hat{k}$$

$$\hat{k} \cdot \hat{n} = 1$$

$$\iint_S (\nabla \times \bar{F})\,\bar{n}\,dS$$

$$= \iint_S (2+r)\sin\theta\, r\,d\theta\,dr$$

$$= \int_0^{\frac{\pi}{2}} \sin\theta\,d\theta \int_0^r \left(2r + r^2\right)dr$$

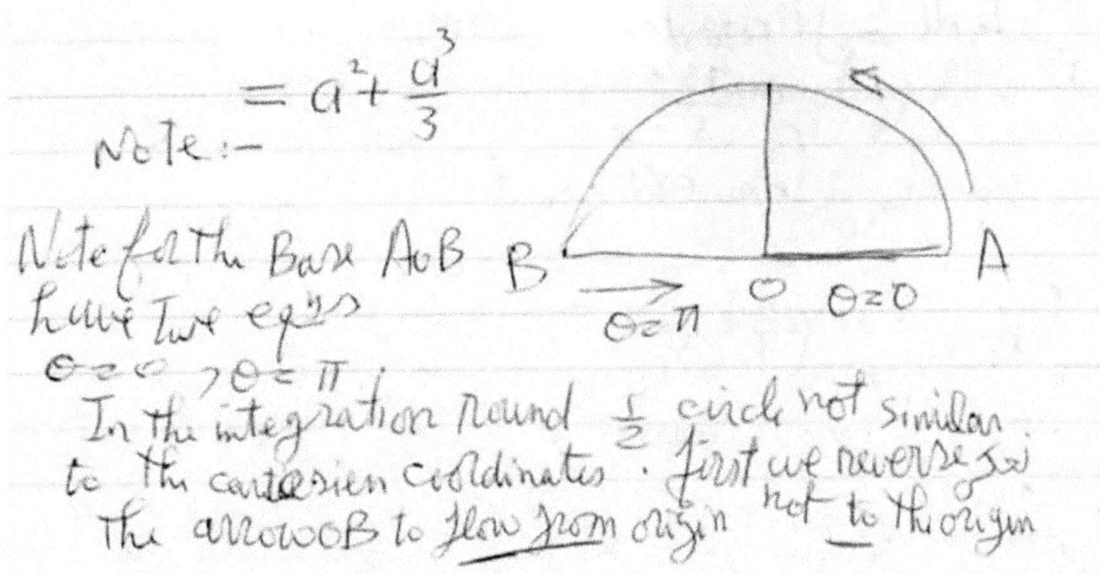

6. Special Orthogonal coordinate Systems

8. Special Orthogonal Coordinate Systems

a) Cylindrical coordinates (ρ, ϕ, z)

$x = \rho \cos \phi, \quad y = \rho \sin \phi \quad z = z,$

where $\rho \geq 0; \quad 0 \leq \phi < 2\pi, \; -\infty < z < \infty.$

$h_\rho = 1, \quad h_\phi = \rho, \; h_z = 1.$

b) Spherical coordinates (r, θ, ϕ)

$x = r \sin \theta \cos \phi, \; y = r \sin \theta \sin \phi, \; z = r \cos \theta,$

where $r \geq 0, \; 0 \leq \theta \leq \pi, \; 0 \leq \phi < 2\pi.$

$h_r = 1, \; h_\theta = r, \; h_\phi = r \sin \theta.$

c) Parabolic cylindrical coordinates (u, v, z)

$x = \frac{1}{2}(u^2 - v^2)$, $y = uv$, $z = z$,

where $-\infty < u < \infty$, $v \geqslant 0$, $-\infty < z < \infty$.

$h_u = h_v = \sqrt{u^2 + v^2}$, $h_z = 1$.

In cylindrical coordinates

$u = \sqrt{2\rho}\cos\frac{\phi}{2}$, $v = \sqrt{2\rho}\sin\frac{\phi}{2}$, $z = z$.

The traces of the coordinate surfaces on the x,y-plane are confocal parabolas with a common axis (Figure 3).

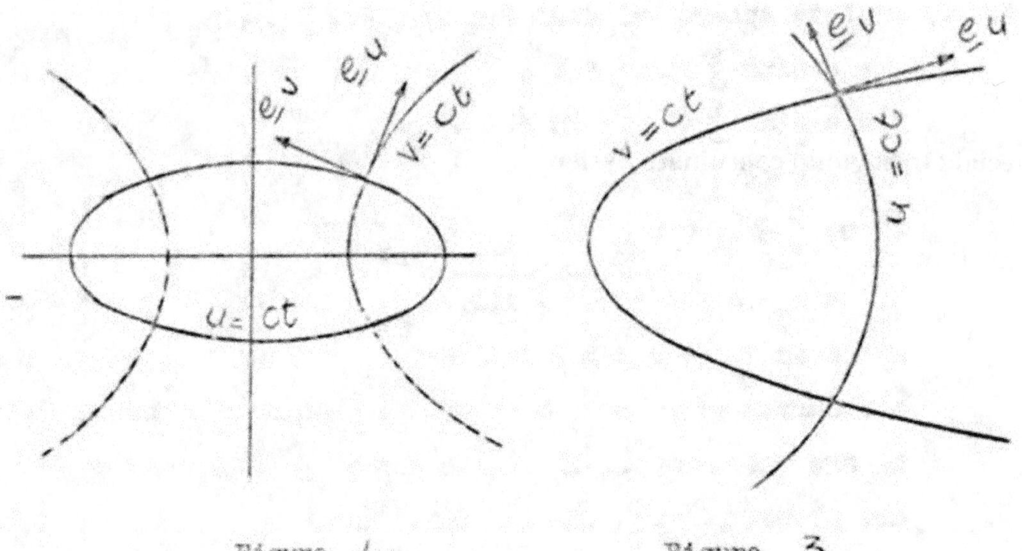

Figure 4.   Figure 3

d) Paraboloidal coordinates (u, v, $\emptyset$)

$x = uv \cos \emptyset, \quad y = uv \sin \emptyset, \quad z = \frac{1}{2}(u^2 - v^2)$

where $u \geq 0, v \geq 0, 0 \leq \emptyset < 2\pi$

$h_u = h_v = \sqrt{u^2 - v^2}, \quad h_\emptyset = uv$

Two sets of coordinate surfaces are obtained by revolving the parabolas of figure 3 above about their axis which is now

the z-axis. The third set of coordinate surfaces are planes through this axis.

e) **Elliptic cylindrical coordinates u, v, z.**

$x = a \cosh u \cos v, \quad y = a \sinh u \sin v, \quad z = z,$

where $u \geq 0, \leq v < 2\pi, -\infty < z < \infty$.

The traces of the coordinate surfaces on the x,y-planes are shown in figure 4. They are confocal ellipses and hyperbolas.

f) Prolate spheroidal coordinates ($\xi, \eta, \emptyset$)

$x = a \sinh \xi \sin \eta \cos \emptyset,$

$y = a \sinh \xi \sin \eta \sin \emptyset,$

$z = a \cosh \xi \cos \eta,$

where $\xi \geq 0, 0 \leq \eta \leq \pi, 0 \leq \emptyset < 2\pi$

$h_\xi = h_\eta = a\sqrt{\sinh^2 \xi + \sin^2 \eta}, \quad h_\emptyset = a \sinh \xi \sin \eta.$

Two sets of coordinate surfaces are obtained by revolving the curves of figure 4 about its x-axis of symmetry which is now the z-axis. The third set of coordinate surfaces are planes passing through this axis.

g) Oblate spheroidal coordinates (ξ, η, ϕ)

$$x = a \cosh \xi \cos \eta \cos \phi, \quad z = a \sinh \xi \sin \eta,$$
$$y = a \cosh \xi \cos \eta \sin \phi,$$

where $\xi \geq 0, -\frac{\pi}{2} \leq \eta \leq \frac{\pi}{2}, 0 \leq \phi < 2\pi$.

$$h_\xi = h_\eta = a\sqrt{\sinh^2 \xi + \sin^2 \eta}, \quad h_\phi = a \cosh \xi \cos \eta.$$

Two sets of coordinate surfaces are obtained by revolving the curves of figure 4 about its y-axis of symmetry which is now the z-axis. The third set of coordinate surfaces are planes passing through this axis.

h) Ellipsoidal coordinates (λ, μ, ν)

$$\frac{x^2}{a^2 - \lambda} + \frac{y^2}{b^2 - \lambda} + \frac{z^2}{c^2 - \lambda} = 1; \; \lambda < c^2 < b^2 < a^2,$$

$$\frac{x^2}{a^2 - \mu} + \frac{y^2}{b^2 - \mu} + \frac{z^2}{c^2 - \mu} = 1; \; c^2 < \mu < b^2 < a^2,$$

$$\frac{x^2}{a^2 - \nu} + \frac{y^2}{b^2 - \nu} + \frac{z^2}{c^2 - \nu} = 1; \; c^2 < b^2 < \nu < a^2.$$

$$h_\lambda = \frac{1}{2}\sqrt{\frac{(\mu - \lambda)(\nu - \lambda)}{(a^2 - \lambda)(b^2 - \lambda)(c^2 - \lambda)}} ,$$

$$h_\mu = \frac{1}{2}\sqrt{\frac{(\nu - \mu)(\lambda - \mu)}{(a^2 - \mu)(b^2 - \mu)(c^2 - \mu)}} ,$$

$$h_\nu = \frac{1}{2}\sqrt{\frac{(\lambda - \nu)(\mu - \nu)}{(a^2 - \nu)(b^2 - \nu)(c^2 - \nu)}} .$$

i) Bipolar coordinates (u, v, z)

$$x^2 + (y - a \cos u)^2 = a^2 \operatorname{cosec}^2 u$$

$$(x - a \coth v)^2 + y^2 = a^2 \operatorname{cosech}^2 v$$

$$z = z$$

or $x = \dfrac{a \sinh v}{\cosh v - \cos u}$, $y = \dfrac{a \sin u}{\cosh v - \cos u}$, $z = z$,

where $0 \leq u < 2\pi$, $-\infty < v < \infty$, $-\infty < z < \infty$.

The traces of the coordinate surfaces on the x,y-plane are shown in figure 5.

By revolving the curves of this figure about its y-axis of symmetry and taking it as z-axis we get a new system of coordinates called the toroidal coordinate system.

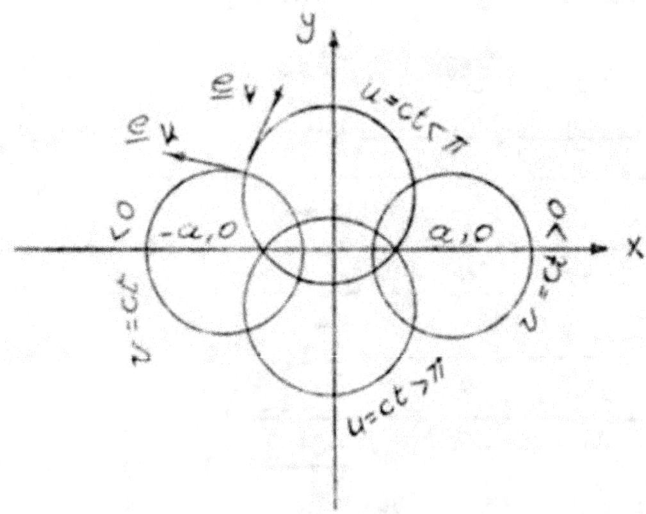

Note: The student is advised to prove the indicated values of h_1, h_2, h_3 and express the functions $\nabla \phi$, $\nabla \cdot \underline{V}$, $\nabla \wedge \underline{V}$, $\nabla^2 \phi$ in each system of coordinates.

7. Examples on Orthogonal Curvilinear Systems

I. a) Prove that a cylindrical coordinate systam is orthogonal.

b) Represent the vector $\underline{A} = z\underline{i} - 2x\,\underline{j} + y\underline{k}$ in cylindrical coordinates.

a) We have $\underline{r} = x\underline{i} + y\underline{j} + z\underline{k} = \rho \cos \phi\ \underline{i} + \rho \sin \phi\ \underline{j} + z\underline{k}$

The tangent to the ρ, ϕ and z curves are given respectively by:

$\dfrac{\partial \underline{r}}{\partial \rho}$, $\dfrac{\partial \underline{r}}{\partial \phi}$ and $\dfrac{\partial \underline{r}}{\partial z}$ where,

$\dfrac{\partial \underline{r}}{\partial \rho} = \cos \phi \underline{i} + \sin \phi\, \underline{j}$, $\dfrac{\partial \underline{r}}{\partial \phi} = -\rho \sin \phi \underline{i} + \rho \cos \phi\ \underline{j}$,

$\dfrac{\partial \underline{r}}{\partial z} = \underline{k}$.

The unit vectors in these directions are

$$\underline{e}_1 = \underline{e}_\rho = \frac{\partial \underline{r}/\partial \rho}{|\partial \underline{r}/\partial \rho|} = \cos \phi \, \underline{i} + \sin \phi \, \underline{j} \, , \quad (1)$$

$$\underline{e}_2 = \underline{e}_\phi = \frac{\partial \underline{r}/\partial \phi}{|\partial \underline{r}/\partial \phi|} = - \sin \phi \, \underline{i} + \cos \phi \, \underline{j} \, , \quad (2)$$

$$\underline{e}_3 = \underline{e}_z = \frac{\partial \underline{r}/\partial z}{|\partial \underline{r}/\partial z|} = \underline{k} \quad . \quad (3)$$

Then $\underline{e}_1 \cdot \underline{e}_2 = 0$, $\underline{e}_1 \cdot \underline{e}_3 = 0$, $\underline{e}_2 \cdot \underline{e}_3 = 0$

and so $\underline{e}_1$, $\underline{e}_2$ and $\underline{e}_3$ are mutually perpendicular and the coordinate system is orthogonal.

b) Solving (1) and (2) simultaneously,

$\underline{i} = \cos \phi \, \underline{e}_\rho - \sin \phi \, \underline{e}_\phi$, $\underline{j} = \sin \phi \, \underline{e}_\rho + \cos \phi \, \underline{e}_\phi$

Then $\underline{A} = z\underline{i} - 2x\underline{j} + y\underline{k}$

$$= (z \cos \phi - 2\rho \cos \phi \sin \phi)\underline{e}_\rho - (z \sin \phi + 2\rho \cos^2 \phi)\underline{e}_\phi + \rho \sin \phi \, \underline{e}_z .$$

II. a) Prove $\frac{d}{dt} \underline{e}_\rho = \dot{\phi} \, \underline{e}_\phi$, $\frac{d}{dt} \underline{e}_\phi = - \dot{\phi} \, \underline{e}_\rho$ where dots denote differentiation with respect to time t

b) Express the velocity $\underline{V}$ and acceleration $\underline{f}$ of a particle in cylindrical coordinates.

a) From (I) we have

$\underline{e}_\rho = \cos \phi \underline{i} + \sin \phi \, \underline{j}$, $\underline{e}_\phi = - \sin \phi \underline{i} + \cos \phi \underline{j}$.

Then $\frac{d}{dt} \underline{e}_\rho = (- \sin \phi \underline{i} + \cos \phi \underline{j}) \dot{\phi} = \dot{\phi} \, \underline{e}_\phi$,

$\frac{d}{dt} \underline{e}_\phi = - (\cos \phi \, \underline{i} + \sin \phi \, \underline{j}) \, \dot{\phi} = - \dot{\phi} \, \underline{e}_\rho$.

b) In cylindrical coordinates using (I) we have

$$\underline{r} = x\underline{i} + y\underline{j} + z\underline{k} = \rho\,\underline{e}_\rho + z\,\underline{e}_z .$$

Then $\underline{V} = \frac{dr}{dt} = \frac{d\rho}{dt}\underline{e}_\rho + \rho\frac{de_\rho}{dt} + \frac{dz}{dt}\underline{e}_z$

$$= \dot\rho\,\underline{e}_\rho + \rho\dot\phi\,\underline{e}_\phi + \dot z\,\underline{e}_z \text{ [using (a)] }.$$

Differentiating again,

$$\underline{f} = \frac{d^2\underline{r}}{dt^2} = \frac{d}{dt}(\dot\rho\,\underline{e}_\rho + \rho\dot\phi\,\underline{e}_\phi + \dot z\,\underline{e}_z)$$

$$= \dot\rho\frac{d\underline{e}_\rho}{dt} + \ddot\rho\,\underline{e}_\rho + \rho\dot\phi\frac{d\underline{e}_\phi}{dt} + \rho\ddot\phi\,\underline{e}_\phi + \dot\rho\dot\phi\underline{e}_\phi + \ddot z\,\underline{e}_z$$

$$= (\ddot\rho - \rho\dot\phi^2)\,\underline{e}_\rho + (\rho\ddot\phi + 2\dot\rho\dot\phi)\underline{e}_\phi + \ddot z\,\underline{e}_z .$$

III. Find the square of the element of arc length for (a) spherical and (b) parabolic cylindrical coordinates.

(a) $x = r\sin\theta\cos\phi,\quad y = r\sin\theta\sin\phi,\ z = r\cos\theta$

Then $dx = -r\sin\theta\sin\phi\,d\phi + r\cos\theta\cos\phi\,d\theta + \sin\theta\cos\phi\,dr,$

$dy = r\sin\theta\cos\phi\,d\phi + r\cos\theta\sin\phi\,d\theta + \sin\theta\sin\phi\,dr,$

$dz = -r\sin\theta\,d\theta + \cos\theta\,dr\ ,$

and $(dS)^2 = (dx)^2 + (dy)^2 + (dz)^2$

$$= (dr)^2 + r^2(d\theta)^2 + r^2\sin^2\theta(d\phi)^2 .$$

The scale factors are $h_1 = h_r = 1,\ h_2 = h_\theta = r,\ h_3 = h_\phi = r\sin\theta .$

(b) $x = \frac{1}{2}(u^2 - v^2),\ y = uv,\ z = z$

Then $dx = u\,du - v\,dv,\qquad dy = u\,dv + v\,du,\qquad dz = dz$

and $(dS)^2 = (dx)^2 + (dy)^2 + (dz)^2$

$$= (u^2+v^2)(du)^2 + (u^2+v^2)(dv)^2 + dz^2$$

The scale factors are

$$h_1 = h_u = \sqrt{u^2+v^2},\qquad h_2 = h_v = \sqrt{u^2+v^2},\qquad h_3 = h_z = 1.$$

IV. Find the volume element $d\tau$ in parabolic cylindrical coordinates.

In parabolic cylindrical coordinates
$u_1 = u$, $u_2 = v$, $u_3 = z$, $h_1 = \sqrt{u^2 + v^2}$,
$h_2 = \sqrt{u^2 + v^2}$, $h_3 = 1$, since

$$(dS)^2 = (u^2+v^2)(du)^2 + (u^2 + v^2)(dv)^2 + (dz)^2].$$

Then $$d\tau = (u^2 + v^2)\, du\, dv\, dz.$$

V. Express in cylindrical coordinates the quantities:

(a) $\nabla \psi$,

(b) $\nabla \cdot \underline{A}$,

(c) $\nabla \wedge \underline{A}$.

For cylindrical coordinates (ρ, ϕ, z),

$$u_1 = \rho, \qquad u_2 = \phi, \qquad u_3 = z;$$

$$\underline{e}_1 = \underline{e}_\rho, \qquad \underline{e}_2 = \underline{e}_\phi, \qquad \underline{e}_3 = \underline{e}_z;$$

and $$h_1 = h_\rho = 1, \; h_2 = h_\phi = \rho, \; h_3 = h_z = 1.$$

(a) $$\nabla \psi = \frac{1}{h_1}\frac{\partial}{\partial u_1}\underline{e}_1 + \frac{1}{h_2}\frac{\partial \psi}{\partial u_2}\underline{e}_2 + \frac{1}{h_3}\frac{\partial \psi}{u_3}\underline{e}_3$$

$$= \frac{\partial \psi}{\partial \rho}\underline{e}_\rho + \frac{1}{\rho}\frac{\partial \psi}{\partial \phi}\underline{e}_\phi + \frac{\partial \psi}{\partial z}\underline{e}_z .$$

(b) $\nabla \cdot \underline{A} = \frac{1}{h_1 h_2 h_3} \left[\frac{\partial}{\partial u_1} (h_2 h_3 A_1) + \frac{\partial}{\partial u_2} (h_3 h_1 A_2) + \frac{\partial}{\partial u_3} (h_1 h_2 A_3) \right]$

$$= \frac{1}{\rho} \left[\frac{\partial}{\partial \rho} (\rho A_\rho) + \frac{\partial A_\phi}{\partial \phi} + \frac{\partial}{\partial z} (\rho A_z) \right]$$

where $\underline{A} = A_\rho \underline{e}_1 + A_\phi \underline{e}_2 + A_z \underline{e}_3$

i.e. $A_1 = A_\rho$, $A_2 = A_\phi$, $A_3 = A_z$

(c) $\nabla \wedge \underline{A} = \frac{1}{h_1 h_2 h_3} \begin{vmatrix} h_1 \underline{e}_1 & h_2 \underline{e}_2 & h_3 \underline{e}_3 \\ \frac{\partial}{\partial u_1} & \frac{\partial}{\partial u_2} & \frac{\partial}{\partial u_3} \\ h_1 A_1 & h_2 A_2 & h_3 A_3 \end{vmatrix}$

$$= \frac{1}{\rho} \begin{vmatrix} \underline{e}_\rho & \rho \underline{e}_\phi & \underline{e}_z \\ \frac{\partial}{\partial \rho} & \frac{\partial}{\partial \phi} & \frac{\partial}{\partial z} \\ A_\rho & \rho A_\phi & A_z \end{vmatrix}$$

$$= \frac{1}{\rho} \Big[\Big(\frac{\partial A_z}{\partial \phi} - \frac{\partial}{\partial z} (\rho A_\phi) \Big) \underline{e}_\rho$$

$$+ \Big(\rho \frac{\partial A_\rho}{\partial z} - \rho \frac{\partial A_z}{\partial \rho} \Big) \underline{e}_\phi$$

$$+ \Big(\frac{\partial}{\partial \rho} (\rho A_\phi) - \frac{\partial A_\rho}{\partial \phi} \Big) \underline{e}_z \Big] .$$

VI. Express Schrödinger's equation of quantum mechanics

$$\nabla^2\psi + \frac{8\pi^2 m}{h^2}[E - V(x,y,z)]\ \psi = 0,$$ where m, h and E are constants.

Here $u_1 = r$, $u_2 = \theta$, $u_3 = \phi$; $h_1 = 1$, $h_2 = r$, $h_3 = r\sin\theta$. From V.

$$\nabla^2\psi = \frac{1}{h_1h_2h_3}\left[\frac{\partial}{\partial u_1}\left(\frac{h_2h_3}{h_1}\frac{\partial\psi}{\partial u_1}\right) + \frac{\partial}{\partial u_2}\left(\frac{h_3h_1}{h_2}\frac{\partial\psi}{\partial u_2}\right) + \frac{\partial}{\partial u_3}\left(\frac{h_1h_2}{h_3}\frac{\partial\psi}{\partial u_3}\right)\right],$$

$$= \frac{1}{(1)(r)(r\sin\theta)}\left[\frac{\partial}{\partial r}\left(\frac{(r)(r\sin\theta)}{(1)}\frac{\partial\psi}{\partial r}\right) + \frac{\partial}{\partial\theta}\left(\frac{(r\sin\theta)(1)}{r}\frac{\partial\psi}{\partial\theta}\right) + \frac{\partial}{\partial\phi}\left(\frac{(1)(r)}{r\sin\theta}\frac{\partial\psi}{\partial\phi}\right)\right]$$

$$= \frac{1}{r^2}\frac{\partial}{\partial r}\left(r^2\frac{\partial\psi}{\partial r}\right) + \frac{1}{r^2\sin\theta}\frac{\partial}{\partial\theta}\left(\sin\theta\frac{\partial\psi}{\partial\theta}\right) + \frac{1}{r^2\sin^2\theta}\frac{\partial^2\psi}{\partial\phi^2}.$$

Then Schrödinger's equation of quantum mechanics is give by

$$\frac{1}{r^2}\frac{\partial}{\partial r}\left(r^2\frac{\partial\psi}{\partial r}\right) + \frac{1}{r^2\sin\theta}\frac{\partial}{\partial\theta}\left(\sin\theta\frac{\partial\psi}{\partial\theta}\right) + \frac{1}{r^2\sin^2\theta}\frac{\partial^2\psi}{\partial\phi^2} + \frac{8\pi^2 m}{h^2}[E - V(r, \theta, \phi)]\psi = 0,$$ where

$\psi = \psi\ (r, \theta, \phi)$.

8. Exercises on Orthogonal Curvilinear Systems

1. Find the unit vectors $\underline{e}_r$, $\underline{e}_\theta$, $\underline{e}_\phi$ of a spherical coordinate system in terms of $\underline{i}$, $\underline{j}$, $\underline{k}$. Solve for $\underline{i}$, $\underline{j}$, $\underline{k}$ in terms of $\underline{e}_r$, $\underline{e}_\theta$, $\underline{e}_\phi$.

   Represent the vector $\underline{A} = -2y\underline{i} - z\underline{j} + 3x\underline{k}$ in spherical coordinates and determine A_r, A_θ and A_ϕ.

2. Represent the vector $\underline{A} = z\underline{i} - 2x\underline{j} + y\underline{k}$ in cylindrical coordinates.

3. Prove that $\dot{\underline{e}}_r = \dot{\theta}\,\underline{e}_\theta + \sin\theta\,\dot{\phi}\,\underline{e}_\phi$, $\dot{\underline{e}}_\theta = -\dot{\theta}\,\underline{e}_r + \cos\theta\,\dot{\phi}\underline{e}_\phi$, $\dot{\underline{e}}_\phi = -\sin\theta\,\dot{\phi}\,\underline{e}_r - \cos\theta\,\dot{\phi}\,\underline{e}_\theta$, where dots denote differentiation with respect to time t, and r, θ, ϕ are the spherical coordinates.

   Express the velocity $\underline{v}$ and the acceleration $\underline{f}$ of a particle in spherical coordinates.

4. Evaluate $\iiint (x^2 + y^2 + z^2)\,dx\,dy\,dz$ over the volume of a sphere of radius a and centre at the origin.

5. Given that $f(r, \theta, \phi) = a\,r^2 \sin\theta \cos^2\phi$ where (r, θ, ϕ) are spherical coordinates, calculate the derivative of f in the direction.

$$\underline{s} = \frac{1}{4}\left(3\,\hat{\underline{r}} + \sqrt{3}\,\hat{\underline{\theta}} + 2\,\hat{\underline{\phi}}\right)$$

   at the point $(a, \frac{\pi}{3}, \frac{\pi}{4})$.

6. Express Maxwell's equation

$$\nabla \wedge \underline{E} = -\frac{1}{c}\frac{\partial \underline{H}}{\partial t}$$

   in prolate spheroidal coordinates.

7. Express Schrœdinger's equation of quantum mechanics

$$\nabla^2 \psi + \frac{8\pi^2 m}{h^2}(E - V(x, y, z)\,\Psi = 0$$

in parabolic cylindrical coordinates, where m, h and E are constants.

8. Express Laplace's equation in spherical polar coordinates r, θ, ϕ. Determine the most general function $f(r)$ such that $v = f(r)\cos\theta$ satisfies this equation.

9. Express the heat equation $\frac{\partial u}{\partial t} = K\nabla^2 u$ in spherical coordinates if u is independent of:

a) ϕ, b) ϕ and θ, c) r and t, d) ϕ, θ and t.

10. Solve problem 8 for cylindrical coordinates.

11. Show that the fundamental equation of elasticity:

$$(\lambda + \mu)\,\text{grad div}\,\underline{D} + \mu\nabla^2\underline{D} = \rho(\underline{f} - \underline{F})$$

can be expressed as:

$$(\lambda + 2\mu)\,\text{grad div}\,\underline{D} - \mu\,\text{curl}\,\underline{D} = \rho\,(\underline{f} - \underline{F})$$

Express the last equation in terms of spherical polar coordinates. When $\underline{D}$, $\underline{F}$ and $\underline{f}$ are radial and functions of r only, show that:

$$(\lambda + 2\mu)\frac{d}{dr}\left(\frac{dD}{dr} + \frac{2D}{r}\right) = \rho\,(f - F)\,.$$

Solve this equation when $f = 0$, $F = -r$ and λ, μ, and ρ are constants.

CHAPTER 4: TENSOR ANALYSIS

1. Spaces of N dimensions

Physical laws must be independent of any particular coordinate systems used in describing them mathematically, if they are to be valid. A study of the consequences of this requirement leads to tensor analysis, of great use in general relativity theory, differential geometry, mechanics, elasticity, hydrodynamics, electromagnetic theory and numerous other fields of science. In three dimensional space a point is a set of three

numbers called coordinates, determined by specifying coordinate system or frame of reference. For example (x, y, z), $(\rho, \emptyset, z)$, $(r, \theta, \emptyset)$ are coordinates of a point in rectangular, cylindrical and spherical coordinates, respectively. A point in N dimensional space is, by analogy, a set of N numbers denoted by $(x^1, x^2, .., x^N)$ where 1, 2, ..., N are taken not as exponents but as superscripts.

2. Coordinate transformations

Let $(x^1, x^2, \ldots, x^N)$ and $(x^{-1}, x^{-2}, \ldots, x^{-N})$ be coordinates of a point in two different frames of references. Suppose there exists N independent relations between the coordinates of the two systems having the form:

$$\left.\begin{array}{l} x^{-1} = x^{-1}(x^1, x^2, \ldots, x^N) \\ x^{-2} = x^{-2}(x^1, x^2, \ldots, x^N) \\ \cdot \quad \ldots\ldots\ldots\ldots\ldots\ldots \\ \ldots\ldots\ldots\ldots\ldots\ldots\ldots \\ \ldots\ldots\ldots\ldots\ldots\ldots\ldots \\ x^{-N} = x^{-N}(x^1, x^2, \ldots, x^N) \end{array}\right\} \qquad (1)$$

which we can indicate briefly by:

$$x^{-k} = x^{-k}(x^1, x^2, \ldots, x^N),\ k = 1, 2, \ldots, N\ , \qquad (2)$$

where it is supposed that the functions involved are single-valued, continuous, and have continuous derivatives. Then conversely to each set of coordinates $(x^{-1}, x^{-2}, \ldots, x^{-N})$ there will correspond a unique set $(x^1, x^2, \ldots, x^N)$given by:

$$x^k = x^k(x^{-1}, x^{-2}, \ldots, x^{-N}),\ k = 1, 2, \ldots, N \qquad (3)$$

The relations (2) or (3) define a transformation of coordinates from one frame of reference to another.

3. The summation convention

In writing an expression such as $a_1x^1 + a_2x^2 + \ldots + a_Nx^N$ we can use the short notation $\sum_{j=1}^{N} a_jx^j$. An even shorter notation is simply to write it as a_jx^j, where we adopt the convention that whenever an index (subscript or superscript) is repeated in a given term we are to sum over that index from 1 to N unless otherwise specified. This is called the summation convention.

Clearly, instead of using the index j, we could have used another letter, say P, and the sum could be written a_px^p. Any index which is repeated in a given term, so that the summation convention applies, is called a dummy index. An index occurring only once in a given term is called a free index and can stand for any of the numbers 1, 2, ..., N such as k in equation (2) or (3), each of which represents N equations.

4. Contra-variant and covariant vectors

If N quantities $A^1, A^2, \ldots, A^N$ in a coordinate system $(x^1, x^2, \ldots, x^N)$ are related to N other quantities $\bar{A}^1, \bar{A}^2, \ldots, \bar{A}^N$) in another coordinate system $(\bar{x}^1, \bar{x}^2, \ldots, \bar{x}^N)$ by the transformation equations

$$A^{-p} = \sum_{q=1}^{N} \frac{\partial x^p}{\partial x^q} A^q, \quad p = 1, 2, \ldots, N,$$

which by the conventions adopted can simply be written as

$$A^{-p} = \frac{\partial x^{-p}}{\partial x^q} A^q,$$

they are called components of a contravariant vector or contravatiant tensor of the first rank or first order. If N quantities $A_1, A_2, \ldots, A_N$ in a coordinate system $(x^1, x^2, \ldots, x^N)$ are related to N other quantities $\bar{A}_1, \bar{A}_2, \ldots, \bar{A}_N$ in another coordinate system $(x^{-1}, x^{-2}, \ldots, x^{-N})$ by the transformation equations.

$$\bar{A}_p = \sum_{q=1}^{N} \frac{\partial x^q}{\partial \bar{x}^p} A_q \quad , \quad p = 1, 2, \ldots, N \, ,$$

or

$$\bar{A}_p = \frac{\partial x^q}{\partial \bar{x}^p} A_q \, , \quad p = 1, 2, \ldots, N,$$

they are called components of a covariant vector or covariant tensor of the first rank or first order.

Note that a superscript is used to indicate contravariant components whereas a subscript is used to indicate covarient components; an exception occurs in the notation for coordinates. Instead of speaking of a tensor whose components are A^p or A_p we shall refer simply to the tensor A^p or A_p.

5. Contravariant, covariant and mixed tensors

If N^2 quantities A^{qs} in a coordinate system $(x^1, x^2, \ldots, x^N)$ are related to N^2 other quantities $\bar{A}^{pr}$ in another coordinate system $(\bar{x}^1, \bar{x}^2, \ldots, \bar{x}^N)$ by the transformation equations

$$\bar{A}^{pr} = \sum_{s=1}^{N} \sum_{q=1}^{N} \frac{\partial \bar{x}^p}{\partial x^q} \frac{\partial \bar{x}^r}{\partial x^s} A^{qs}, \quad p, r = 1, 2, \ldots, N$$

or

$$\bar{A}^{pr} = \frac{\partial \bar{x}^p}{\partial x^q} \frac{\partial \bar{x}^r}{\partial x^s} A^{qs} ,$$

by the adopted conventions, they are called contravariant components of a tensor of the second rank or rank two. The N^2 quantities A_{qs} are called covariant components of a tensor of the second rank if

$$\bar{A}_{pr} = \frac{\partial x^q}{\partial \bar{x}^p} \frac{\partial x^s}{\partial \bar{x}^r} A_{qs} .$$

Similarly the N^2 quantities A^q_s are called components of a mixed tensor of the second rank if

$$\bar{A}^p_r = \frac{\partial \bar{x}^p}{\partial x^q} \frac{\partial x^s}{\partial \bar{x}^r} A^q_s .$$

by the adopted conventions, they are called contravariant components of a tensor of the second rank or rank two. The N^2 quantities A_{qs} are called covariant components of a tensor of the second rank if

$$\bar{A}_{pr} = \frac{\partial x^q}{\partial \bar{x}^p}\frac{\partial x^s}{\partial \bar{x}^r} A_{qs} .$$

Similarly the N^2 quantities A^q_s are called components of a mixed tensor of the second rank if

$$\bar{A}^p_r = \frac{\partial \bar{x}^p}{\partial x^q}\frac{\partial x^s}{\partial \bar{x}^r} A^q_s .$$

6. The Kronecker delta

The Kronecker delta, written δ^j_k, is defined by:

$$\delta^j_k = \begin{cases} 0 & \text{if } j \neq k , \\ 1 & \text{if } j = k. \end{cases}$$

As its notation indicates, it is a mixed tensor of the second rank.

7. Tensor of rank greater than two

They are easily defined. For example $A^{qst}_{k\ell}$ are the component of a mixed tensor of rank 5, contravatient of order 3 and covariant of order 2, if they transform according to the relations.

$$\bar{A}^{prm}_{ij} = \frac{\partial \bar{x}^p}{\partial x^q}\frac{\partial \bar{x}^r}{\partial x^s}\frac{\partial \bar{x}^m}{\partial x^t}\frac{\partial x^k}{\partial \bar{x}^i}\frac{\partial x^\ell}{\partial \bar{x}^j} A^{qst}_{k\ell}$$

8. Scalars or invariants and tensor fields

Suppose ϕ is a function of the coordinates x^k, and let $\bar{\phi}$ denote the functional value under a transformation to a new set of coordinates $\bar{x}^k$. Then ϕ is called a scalar or invariant with respect to the coordinates transformation if $\phi = \bar{\phi}$. A scalar or invariant is also called a tensor of rank zero.

If to each point of a region in N dimensional space there corresponds a definite tensor, we say that a tensor field has been defined. This is a vector field or a scalar field according as the tensor is of rank one or zero. It should be noted that a tensor or tensor field is not just the set of its components in one special coordinate system but all the possible sets under any transformation of coordinates.

9. Symmetric and skew—symmetric tensors

A tesnor is called symmetric with respect to two contravariant or two covariant indices if its components remain unaltered upon interchange of the indicies. Thus if $A^{mpr}_{qs} = A^{pmr}_{qs}$ the tensor is symmetric in m and p. If a tensor is symmetric with respect to any two contravariant and any two covariant indices, it is called symmetric.

A tensor is called skew-symmetric with respect to two contravariant or two covariant indices, if its components change sign upon interchange of the indices. Thus if $A^{mpr}_{qs} = - A^{pmr}_{qs}$, the tensor is skew-symmetric in m and p. If a tensor is skew-symmetric with respect to any two contraviant and any two covariant indices, it is called skew-symmetric.

10. Examples

i. Write the law of transformation for the tensor.

(a) A^i_{jk} , (b) B^{mn}_{ijk} , (c) C^m

(b) $\bar{A}^p_{qr} = \dfrac{\partial \bar{x}^p}{\partial x^i} \dfrac{\partial x^j}{\partial \bar{x}^q} \dfrac{\partial x^k}{\partial \bar{x}^r} A^i_{jk}$,

(c) $\bar{B}^{pq}_{rs} = \dfrac{\partial \bar{x}^p}{\partial x^m} \dfrac{\partial \bar{x}^q}{\partial x^n} \dfrac{\partial x^i}{\partial \bar{x}^r} \dfrac{\partial x^j}{\partial \bar{x}^s} \dfrac{\partial x^k}{\partial \bar{x}^t} B^{mn}_{ijk}$,

(d) $\bar{C}^p = \dfrac{\partial \bar{x}^p}{\partial x^m} C^m$.

ii. A covariant tensor has components xy, $2y - z^2$, xz in rectangular coordinates. Find its covariant components in spherical coordinates.

Let A_j denote the covariant components in rectangular coordinates $x^1 = x$, $x^2 = y$, $x^3 = z$. Then

$A_1 = xy = x^1x^2$, $A_2 = 2y - z^2 = 2x^2 - (x^3)^2$, $A_3 = x^1x^3$

where care must be taken to distinguish between superscripts and exponents.

Let $\bar{A}_k$ denote the covariant components in spherical coordinates:

$\bar{x}^1 = r$, $x^{-2} = \theta$, $\bar{x}^3 = \phi$. Then

$$\bar{A}_k = \frac{\partial x^j}{\partial \bar{x}^k} A_j \; . \qquad (1)$$

The transformation equations between coordinate systems are

$$x^1 = \bar{x}^1 \sin \bar{x}^2 \cos \bar{x}^3, \; x^2 = \bar{x}^1 \sin \bar{x}^2 \sin \bar{x}^3, \; x^3 = \bar{x}^1 \cos \bar{x}^2$$

Then equations (1) yield the required covariant components.

$$\bar{A}_1 = \frac{\partial x^1}{\partial \bar{x}^1} A_1 + \frac{\partial x^2}{\partial \bar{x}^1} A_2 + \frac{\partial x^3}{\partial \bar{x}^1} A_3$$

$$= (\sin \bar{x}^2 \cos \bar{x}^3)(x^1x^2) + (\sin \bar{x}^2 \sin \bar{x}^3)(2x^2 - (x^3)^2)$$

$$+ (\cos \bar{x}^2)(x^1x^3)$$

$$= (\sin \theta \cos \phi)\;(r^2 \sin^2 \theta \sin \phi \cos \phi)$$

$$+ (\sin \theta \sin \phi)(2r \sin \theta \sin \phi - r^2 \cos^2 \theta)$$

$$+ (\cos \theta)(r^2 \sin \theta \cos \theta \cos \phi) \; .$$

By the same way we can obtain $\bar{A}_2$, $\bar{A}_3$.

iii. (a) Show that $\dfrac{\partial x^p}{\partial x^q} = \delta^p_q$,

(b) Prove that $\dfrac{\partial x^p}{\partial \bar{x}^q} \dfrac{\partial \bar{x}^q}{\partial x^r} = \delta^p_r$

(a) If $p \neq q$, $\dfrac{\partial x^p}{\partial x^q} = 0$ since x^p and x^q are independent.

If $p = q$ then $\dfrac{\partial x^p}{\partial x^q} = 1$, since $x^p = x^q$. Therefore $\dfrac{\partial x^p}{\partial x^q} = \delta^p_q$.

(b) Coordinates x^p are functions of coordinates $\bar{x}^q$ which are in turn functions of coordinates x^r. Then by the chain rule and example (iii"a") we have,

$$\frac{\partial x^p}{\partial x^r} = \frac{\partial x^p}{\partial \bar{x}^q} \frac{\partial \bar{x}^q}{\partial x^r} = \delta^p_r$$

IV. If $\phi = a_{jk} A^j A^k$, show that we can always write

$$\phi = b_{jk} A^j A^k \text{ where } b_{jk} \text{ is symmetric.}$$

$$\phi = a_{jk} A^j A^k = a_{kj} A^k A^j \quad = a_{kj} A^j A^k.$$

Then $$2\phi = a_{jk} A^j A^k + a_{kj} A^j A^k = (a_{jk} + a_{kj}) A^j A^k,$$

and $$\phi = \tfrac{1}{2} (a_{jk} + a_{kj}) A^j A^k = b_{jk} A^j A^k,$$

where $$b_{jk} = \tfrac{1}{2} (a_{jk} + a_{kj}) = b_{kj} \text{ is symmetric.}$$

11. Fundamental operations with tensors

(a) Addition and subtraction

The sum of two or more tensors of the same rank and type (i.e. same number of contravariant indices and same number of covariant indices) is also a tensor of the same rank and type. Thus if A^{mp}_q and B^{mp}_q are tensors, then $C^{mp}_q = A^{mp}_q + B^{mp}_q$ is also a tensor. Addition of tensors is commutative and associative.

The difference of two tensors of the same rank and type is also a tensor of the same rank and type. Thus $D^{mp}_q = A^{mp}_q - B^{mp}_q$ is also a tensor.

(b) Outer multiplication

The product of two tensors is a tensor whose rank is the sum of the ranks of the given tensors. This product which involves ordinary multiplication of the components of the tensor is called the outer product. For example $A^{pr}_{q}\ B^{m}_{s} = C^{prm}_{qs}$ is the outer product of A^{pr}_{q} and B^{m}_{s}. However, note that not every tensor can be written as a product of two tensors of lower rank. For this reason division of tensors is not always possible.

(c) Contraction

If one contravariant and one covariant index of a tensor are set equal, the result indicates that a summation over the equal indices is to be taken according to the summation convention. This resulting sum is a tensor of rank two less than that of the original tensor. The process is called contraction. For example, in the tensor of rank 5, A^{mpr}_{qs}, set r = s to obtain $A^{mpr}_{qr} = B^{mp}_{q}$ a tensor of rank 3. Further, by setting p = q we obtain $B^{mp}_{q} = C^{m}$ a tensor of rank 1.

(d) Inner multiplication

By the process of outer multiplication of two tensors followed by a contraction, we obtain a new tensor called inner multiplication. For example, given the tensors A^{mp}_{q} and B^{r}_{st}, the outer product is $A^{mp}_{q}\ B^{r}_{st}$. Letting q = r, we obtain the inner product $A^{mp}_{r}\ B^{r}_{st}$. Letting q = r and p = s, another inner product $A^{mp}_{r}\ B^{r}_{pt}$ is obtained. Inner and outer multiplication of tensors is commutative and associative.

(e) Quotient law

Suppose it is not known whether a quantity X is a tensor or not. If an inner product of X with an arbitrary tensor is itself a tensor, then X is also a tensor. This is called the quotient law.

12. Examples on tensor calculus

1. If A_r^{pq} and B_t^s are tensors, prove that $C_{rt}^{pqs} = A_r^{pq}\, B_t^s$ is also a tensor.

We must prove that C_{rt}^{pqs} is a tensor whose components are formed by taking the products of components of tensors A_r^{pq} and B_t^s. Since A_r^{pq} and B_t^s are tensors,

$$\bar{A}_\ell^{jk} = \frac{\partial \bar{x}^j}{\partial x^p} \frac{\partial \bar{x}^k}{\partial x^q} \frac{\partial x^r}{\partial \bar{x}^\ell} A_r^{pq},$$

$$\bar{B}_n^m = \frac{\partial \bar{x}^m}{\partial x^s} \frac{\partial x^t}{\partial \bar{x}^n} B_t^s .$$

Multiplying,

$$\bar{A}_\ell^{jk}\, \bar{B}_n^m = \frac{\partial \bar{x}^j}{\partial x^p} \frac{\partial \bar{x}^k}{\partial x^q} \frac{\partial x^r}{\partial \bar{x}^\ell} \frac{\partial \bar{x}^m}{\partial x^s} \frac{\partial x^t}{\partial \bar{x}^n} A_r^{pq}\, B_t^s$$

which shows that $A_r^{pq}\, B_t^s$ is a tensor of rank 5, with contravariant indices p, q, s and covariant indices r,t, thus warranting the notation C_{rt}^{pqs}. We call $C_{rt}^{pqs} = A_r^{pq}\, B_t^s$ the outer product of A_r^{pq} and B_t^s.

ii. Show that any inner product of the tensor A_r^p and B_t^{qs} is a tensor of rank three.

Outer product of A_r^p and $B_t^{qs} = A_r^p\, B_t^{qs}$.

Let us contract with respect to indices p and t, i.e. let p = t and sum. We must show that the resulting inner product, represented by $A_r^p\, B_t^{qs}$, is a tensor of rank three.

By hypothesis, A_r^p and B_t^{qs} are tensors; then

$$\bar{A}_k^j = \frac{\partial \bar{x}^j}{\partial x^p} \frac{\partial x^r}{\partial \bar{x}^k} A_r^p ,$$

$$\bar{B}_n^{lm} = \frac{\partial \bar{x}^l}{\partial x^q} \frac{\partial \bar{x}^m}{\partial x^s} \frac{\partial x^t}{\partial \bar{x}^n} \quad B_t^{qs}$$

Multiplying, letting j =n and summing, we have:

model closer to nature than the Newtonian model. We must therefore pay attention to the words: There is no such thing as the time, in any absolute sense. Once that point is conceded, the basis of the Newtonian pattern is broken and the way is opçn for relativity.

The theory of relativity is divided into two parts:

(i) The special theory,

(ii) The general theory.

The special theory deals with phenomena in which gravitational attraction plays no part. While the general theory might be called "Einstein's theory of gravitation". We shall be concerned with the special theory.

First we introduce a particle, next we introduce a frame of reference and an observer in it. The observer has a measuring rod with which he can measure the distances between the particles which form his frame of reference. If the distances between these particles remain constant, the observer declares that his frame of reference is a rigid body. Now we provide

the observer with a clock. As we have indicated above, the transporting of a clock is an operation which may lead to curious consequences. We shall therefore not expect the observer to carry his clock about but shall provide him with a great number of clocks, all of identical construction. These will be distributed throughout his frame of reference and kept

$$\bar{A}^j_k \bar{B}^{\ell m}_j = \frac{\partial \bar{x}^j}{\partial x^p} \frac{\partial x^r}{\partial \bar{x}^k} \frac{\partial \bar{x}^\ell}{\partial x^q} \frac{\partial \bar{x}^m}{\partial x^s} \frac{\partial x^t}{\partial \bar{x}^s} A^p_r B^{qs}_t$$

$$= \delta^t_p \frac{\partial x^r}{\partial \bar{x}^k} \frac{\partial \bar{x}^\ell}{\partial x^q} \frac{\partial \bar{x}^m}{\partial x^s} A^p_r B^{qs}_t$$

$$= \frac{\partial x^r}{\partial \bar{x}^k} \frac{\partial \bar{x}^\ell}{\partial x^q} \frac{\partial \bar{x}^m}{\partial x^s} A^p_r B^{qs}_p .$$

Showing that $A^p_r B^{qs}_p$ is a tensor of rank three. By contracting with respect to q and r or s and r in the product $A^p_r B^{qs}_t$. We can similarly show that any inner product is a tensor of rank three.

13. The line element and metric tensor

We have $$d\underline{r} = \frac{\partial \underline{r}}{\partial u_1} du_1 + \frac{\partial \underline{r}}{\partial u_2} du_2 + \frac{\partial \underline{r}}{\partial u_3} du_3$$

$$= \underline{\alpha}_1 du_1 + \underline{\alpha}_2 du_2 + \underline{\alpha}_3 du_3$$

$$\therefore (ds)^2 = d\underline{r} \cdot d\underline{r}$$

$$= \underline{\alpha}_1 \cdot \underline{\alpha}_1 du_1^2 + \underline{\alpha}_1 \cdot \underline{\alpha}_2 du_1 du_2 + \underline{\alpha}_1 \cdot \underline{\alpha}_3 du_1 du_3$$

$$+ \underline{\alpha}_2 \cdot \underline{\alpha}_1 du_2 du_1 + \underline{\alpha}_2 \cdot \underline{\alpha}_2 du_2^2 + \underline{\alpha}_2 \cdot \underline{\alpha}_3 du_2 du_3$$

$$+ \underline{\alpha}_3 \cdot \underline{\alpha}_1 du_3 du_1 + \underline{\alpha}_3 \cdot \underline{\alpha}_2 du_3 du_2 + \underline{\alpha}_3 \cdot \underline{\alpha}_3 du_3^2$$

$$= \sum_{p=1}^{3} \sum_{q=1}^{3} g_{pq} du_p du_q \quad \text{where} \quad g_{pq} = \underline{\alpha}_p \cdot \underline{\alpha}_q$$

This is called the fundamental quadratic form or metric form. The quantities g_{pq} are called metric coefficients and are symmetric. Such spaces are called three dimensional Euclidean spaces.

A generalization to N dimensional space with coordinates $(x^1, x^2, \ldots, x^N)$ is immediate. We define the line element dS in this space to be given by the quadratic form, called the metric form or metric,

$$dS^2 = \sum_{p=1}^{N} \sum_{q=1}^{N} g_{pq} \, dx^p \, dx^q ,$$

or, using the summation convention,

$$dS^2 = g_{pq} \, dx^p dx^q$$

In the special case where there exists a transformation of coordinates from x^j to $\bar{x}^k$ such that the metric form is transformed into $(d\bar{x}^1)^2 + (d\bar{x}^2)^2 + \ldots (d\bar{x}^N)^2$ or $d\bar{x}^k\, d\bar{x}^k$, then the space is called N-dimensional Euclidean space. In the general case, however, the space is called Riemannian.

The quantities g_{pq} are the components of a covariant tensor of rank two called the metric tensor or fundamental tensor. We can and always will choose this tensor to be symmetric.

14. Conjugate or reciprocal tensors

Let $g = g_{pq}$ denote the determinant with elements g_{pq} and suppose $g \neq 0$. Define g^{pq} by

$$g^{pq} = \frac{\text{cofactor of } g_{pq}}{g}\,.$$

(g — determinant whose elements are g_{pq})

Then g^{pq} is a symmetric contravariant tensor of rank two called the conjugate or reciprocal tensor of g_{pq}. It can be shown that

$$g^{pq}\, g_{rq} = \delta^p_r\,.$$

15. Associated tensors

Given a tensor, we can derive other tensors by raising or lowering indices. For example, given the tensor A_{pq} we obtain by raising the index p, the tensor $A^{p}_{\cdot q}$, the dot indicating the original position of the pq moved index. By raising the index q also we obtain $A^{pq}_{\cdot\cdot}$. Where no confusion can arise we shall often omit the dots; thus $A^{pq}_{\cdot\cdot}$ can be written A^{pq}. These derived tensors can be obtained by forming inner products of the given tensor with the metric tensor g_{pq} or its conjugate g^{pq}. Thus for example:

$$A^{p}_{\cdot q} = g^{rp} A_{rq} \,, \quad A^{p}_{\cdot rs} = g_{rq} A^{pq}_{\cdot\cdot s}$$

$$A^{qm\cdot tk}_{\cdot\cdot n} = g^{pk} g_{sn} g^{rm} A^{q\cdot st}_{\cdot r\cdot\cdot p}$$

These become clear if we interpret multiplication by g^{rp} as meaning: let r = p (or p = r) in whatever follows and raise this index. Similarly we interpret multiplication by g_{rq} as meaning: let r = q (or q = r) in whatever follows and lower this index.

All tensors obtained from a given tesnor by forming inner products with the metric tensor and its conjugate are called associated tensors of the given tensor. For example A^{m} and A_{m} are associated tensors, the first are contravariant and the second covariant components. The relation between them is given by:

$$A_{p} = g_{pq} A^{q} \quad \text{or} \quad A^{p} = g^{pq} A_{q}$$

For rectangular coordinates $g_{pq} = 1$, if $p = q$, and 0 if $p \neq q$, so that $A_p = A^p$, which explains why no distinction was made between contravariant and covariant components of a vector in earlier chapters.

16. Length of a vector, angle between vectors

The quantity A^pB_p, which is the inner product of A^p and B_q, is a scalar analogous to the scalar product in rectangular coordinates. We define the length L of the vector A^p or A_p as given by:

$$L^2 = A^pA_p = g^{pq} A_pA_q = g_{pq} A^pA^q .$$

We can define the angle θ between A^p and B_p as given by

$$\cos\theta = \frac{A^pB_p}{\sqrt{(A^pA_p)(B^pB_p)}} .$$

17. The physical components of a vector

The physical components of a vector A^p or A_p, denoted by A_u, A_v, and A_w are the projections of the vector on the tangents to the coordinate curves and are given in the case of orthogonal coordinates by:

$$A_u = \sqrt{g_{11}}\ A^1 = \frac{A_1}{\sqrt{g_{11}}}\ ,\quad A_v = \sqrt{g_{22}}\ A^2 = \frac{A_2}{\sqrt{g_{22}}}\ ,$$

$$A_w = \sqrt{g_{33}}\ A^3 = \frac{A_3}{\sqrt{g_{33}}}\ .$$

Similarly, the physical components of a tensor A^{pq} or A_{pq} are given by:

$$A_{uu} = g_{11}\ A^{11} = \frac{A_{11}}{g_{11}}\ ,\quad A_{uv} = \sqrt{g_{11}g_{22}}\ A^{12} = \frac{A_{12}}{\sqrt{g_{11}\ g_{22}}}\ ,$$

$$A_{uw} = \sqrt{g_{11}\ g_{33}}\ A^{13} = \frac{A_{13}}{\sqrt{g_{11}\ g_{33}}}\ ,\ \text{etc.}$$

18. Christoffel's symbols

The symbols

$$[pq,r] = \frac{1}{2}\left(\frac{\partial g_{pr}}{\partial x^q} + \frac{\partial g_{qr}}{\partial x^p} - \frac{\partial g_{pq}}{\partial x^r}\right)$$

$$\left\{ {s \atop pq} \right\} = g^{sr}\ [pq,r]$$

are called the Christoffel symbols of the first and second kind respectively. Other symbols used instead of $\left\{ {s \atop pq} \right\}$ are $\{pq,s\}$ and Γ^s_{pq}. The latter symbol suggests however a tensor character, which is not true in general.

19. Transformation laws of Christoffel's symbols

If we denote by a bar a symbol in a coordinate system $\overline{x}^k$, then:

$$\overline{[jk,m]} = [pq,r] \frac{\partial x^p}{\partial \overline{x}^j} \frac{\partial x^q}{\partial \overline{x}^k} \frac{\partial x^q}{\partial \overline{x}^k} \frac{\partial x^r}{\partial \overline{x}^m} + g_{pq} \frac{\partial x^p}{\partial \overline{x}^m} \frac{\partial^2 x^q}{\partial \overline{x}^j \partial \overline{x}^k} ,$$

$$\overline{\left\{ {n \atop jk} \right\}} = \left\{ {s \atop pq} \right\} \frac{\partial \overline{x}^n}{\partial x^s} \frac{\partial x^p}{\partial \overline{x}^j} \frac{\partial x^q}{\partial \overline{x}^k} + \frac{\partial \overline{x}^n}{\partial x^q} \frac{\partial^2 x^q}{\partial \overline{x}^j \partial \overline{x}^k}$$

are the laws of transformation of the Christoffel symbols showing that they are not tensors unless the second terms on the right are zero.

20. Geodescis

The distance s between two points t_1 and t_2 on a curve $x^r = x^r(t)$ in a Riemannian space is given by

$$S = \int_{t_1}^{t_2} \sqrt{g_{pq} \frac{dx^p}{dt} \frac{dx^q}{dt}} \, dt .$$

That curve in the space which makes the distance a minimum is called a geodesic of the space. By use of the calculus of variations the geodesics are found from the differential equation

$$\frac{d^2 x^r}{dS^2} + \left\{ {r \atop pq} \right\} \frac{dx^p}{dS} \frac{dx^q}{dS} = 0$$

where S is the arc length parameter. As examples, the geodesics on a plane are straight lines whereas the geodesics on a sphere are arcs of great circles.

21. The covariant derivative

The covariant derivative of a tensor A_p with respect to x^q is denoted by $A_{p,q}$ and is defined by.

$$A_{p,q} = \frac{\partial A_p}{\partial x^q} - \begin{Bmatrix} s \\ pq \end{Bmatrix} A_s ,$$

a covariant tensor of rank two.

The covariant derivative of a tensor A^p with respect to x^q is denoted by $A^p_{\ ,q}$ and is defined by

$$A^p_{\ ,q} = \frac{\partial A^p}{\partial x^q} + \begin{Bmatrix} p \\ qs \end{Bmatrix} A^s$$

a mixed tensor of rank two.

For rectangular systems, the Christoffel symbols are zero and the covariant derivative are the usual partial derivative. Covariant derivatives of tensors are also tensors.

The above results can be extended to covariant derivatives of higher rank tensors. Thus

$$A^{p_1 \cdots p_m}_{r_1 \cdots r_{n,q}} = \frac{\partial A^{p_1 \cdots p_m}_{r_1 \cdots r_n}}{\partial x^q}$$

$$- \begin{Bmatrix} s \\ r_1 q \end{Bmatrix} A^{p_1 \cdots p_m}_{s r_2 \cdots r_n} - \begin{Bmatrix} s \\ r_2 q \end{Bmatrix} A^{p_1 \cdots p_m}_{r_1 s r_3 \cdots r_n}$$

$$- \cdots - \begin{Bmatrix} s \\ r_n q \end{Bmatrix} A^{p_1 \cdots p_m}_{r_1 \cdots r_{n-1} s}$$

$$+\begin{Bmatrix} p_1 \\ qs \end{Bmatrix} A^{sp_2 \ldots p_m}_{r_1 \ldots r_n} + \begin{Bmatrix} p_2 \\ qs \end{Bmatrix} A^{p_1 s p_3 \ldots p_m}_{r_1 \ldots r_n}$$

$$+\begin{Bmatrix} p_m \\ qs \end{Bmatrix} A^{p_1 \ldots p_{m-1} s}_{r_1 \ldots r_n}$$

is the covariant derivative of $A^{p_1 \ldots p_m}_{r_1 \ldots r_n}$ with respect to x^q.

22. Permutation symbols and tensors

Further, let us define:

$$\epsilon_{pqr} = \frac{1}{\sqrt{g}} e_{pqr} , \quad \epsilon^{pqr} = \sqrt{g}\, e^{pqr} .$$

It can be shown that ϵ_{pqr} and ϵ^{pqr} are covariant and contravariant tensors respectively, called permutation tensors in three dimensional space. Generalizations to higher dimensions are possible.

Define e_{pqr} by the relations:

$$e_{123} = e_{231} = e_{312} = 1, \; e_{213} = e_{132} = e_{321} = -1. \; e_{pqr} = 0$$

if two or more indices are equal and define e^{pqr} in the same manner. The symbols e_{pqr} and e^{pqr} are called permutation symbols in three dimensional space.

23. Examples on tensor operations

i) Express in matrix notation the transformation equation for (a) a covariant vector, (b) a contravariant tensor of rank two, assuming N = 3.

(a) The transformation equation $\bar{A}_p = \frac{\partial x^q}{\partial \bar{x}^p} A_q$ can be written

$$\begin{pmatrix} \bar{A}_1 \\ \bar{A}_2 \\ \bar{A}_3 \end{pmatrix} \begin{pmatrix} \frac{\partial x^1}{\partial \bar{x}^1} & \frac{\partial x^1}{\partial \bar{x}^2} & \frac{\partial x^1}{\partial \bar{x}^3} \\ \frac{\partial x^2}{\partial \bar{x}^1} & \frac{\partial x^2}{\partial \bar{x}^2} & \frac{\partial x^2}{\partial \bar{x}^3} \\ \frac{\partial x^3}{\partial \bar{x}^1} & \frac{\partial x^3}{\partial \bar{x}^2} & \frac{\partial x^3}{\partial \bar{x}^3} \end{pmatrix} \begin{pmatrix} A_1 \\ A_2 \\ A_3 \end{pmatrix}$$

(b) The transformation equations $\bar{A}^{pr} = \frac{\partial \bar{x}^p}{\partial x^q} \frac{\partial \bar{x}^r}{\partial x^s} A^{qs}$ can be written:

$$\begin{pmatrix} \bar{A}^{11} & \bar{A}^{12} & \bar{A}^{13} \\ \bar{A}^{21} & \bar{A}^{22} & \bar{A}^{23} \\ \bar{A}^{31} & \bar{A}^{32} & \bar{A}^{33} \end{pmatrix} = \begin{pmatrix} \frac{\partial \bar{x}^1}{\partial x^1} & \frac{\partial \bar{x}^1}{\partial x^2} & \frac{\partial \bar{x}^1}{\partial x^3} \\ \frac{\partial \bar{x}^2}{\partial x^1} & \frac{\partial \bar{x}^2}{\partial x^2} & \frac{\partial \bar{x}^2}{\partial x^3} \\ \frac{\partial \bar{x}^3}{\partial x^1} & \frac{\partial \bar{x}^3}{\partial x^2} & \frac{\partial \bar{x}^3}{\partial x^3} \end{pmatrix} \begin{pmatrix} A^{11} & A^{12} & A^{13} \\ A^{21} & A^{22} & A^{23} \\ A^{31} & A^{32} & A^{33} \end{pmatrix} \begin{pmatrix} \frac{\partial \bar{x}^1}{\partial x^1} & \frac{\partial \bar{x}^2}{\partial x^1} & \frac{\partial \bar{x}^3}{\partial x^1} \\ \frac{\partial \bar{x}^1}{\partial x^2} & \frac{\partial \bar{x}^2}{\partial x^2} & \frac{\partial \bar{x}^3}{\partial x^2} \\ \frac{\partial \bar{x}^1}{\partial x^3} & \frac{\partial \bar{x}^2}{\partial x^3} & \frac{\partial \bar{x}^3}{\partial x^3} \end{pmatrix}$$

ii) Determine the metric tensor in (a) cylindrical and (b) spherical coordinates,

(a) $dS^2 = d\rho^2 + \rho^2 \, d\phi^2 + dz^2$.

If $x^1 = \rho$, $x^2 = \phi$, $x^3 = z$, then $g_{11} = 1$, $g_{22} = \rho^2$, $g_{33} = 1$

$g_{12} = g_{21} = 0$, $g_{23} = g_{32} = 0$, $g_{31} = g_{13} = 0$.

(b) $dS^2 = dr^2 + r^2 d\theta^2 + r^2 \sin^2\theta \, d\phi^2$.

If $x^1 = r$, $x^2 = \theta$, $x^3 = \phi$ the metric tensor can be written:

$$\begin{pmatrix} 1 & 0 & 0 \\ 0 & r^2 & 0 \\ 0 & 0 & r^2\sin^2\theta \end{pmatrix}$$

In general for orthogonal coordinate, $g_{jk} = 0$ for $j \neq k$.

iii. Find (a) g and (b) g^{jk} corresponding to

$dS^2 = 5(dx^1)^2 + 3(dx^2)^2 + 4(dx^3)^2 - 6dx^1dx^2 + 4dx^2dx^3$.

(a) $g_{11} = 5$, $g_{22} = 3$, $g_{33} = 4$, $g_{12} = g_{21} = -3$, $g_{23} = g_{32} = 2$,

$g_{13} = g_{31} = 0$. Then

$$g = \begin{vmatrix} 5 & -3 & 0 \\ -3 & 3 & 2 \\ 0 & 2 & 4 \end{vmatrix} = 4$$

(b) The cofactors $G(j,k)$ of g_{ik} are

$G(1,1) = 8$, $G(2,2) = 20$, $G(3,3) = 6$, $G(1,2) = G(2,1) = 12$

$G(2,3) = G(3,2) = -10$, $G(1,3) = G(3,1) = -6$.

Then $g^{11} = 2$, $g^{22} = 5$, $g^{33} = \frac{3}{2}$, $g^{12} = g^{21} = 3$, $g^{23} = g^{32} = -\frac{5}{2}$,

$g^{13} = g^{31} = -\frac{3}{2}$.

Note that the product of the matrices (g_{jk}) and (g^{jk}) is the unit matrix I.

iv. Prove: (a) $[pq,r] = g_{rs} \left\{ {s \atop pq} \right\}$,

(b) $$\frac{\partial g_{pq}}{\partial x^m} = [pm,q] + [qm,p]$$

(c) $$\frac{\partial g^{pq}}{\partial x^m} = - g^{pn} \left\{ {q \atop mn} \right\} - g^{qn} \left\{ {p \atop mn} \right\}.$$

(a) $$g_{ks} \left\{ {s \atop pq} \right\} = g_{ks}\, g^{sr} [pq,r] = \delta^r_k [pq,r] = [pq,r] = [pq,k]$$

or

$$[pq,k] = g_{ks} \left\{ {s \atop pq} \right\} \quad \text{i.e. } [pq,r] = g_{rs} \left\{ {s \atop pq} \right\}.$$

(b) $$[pm,q] + [qm,p] = \frac{1}{2} \left(\frac{\partial g_{pq}}{\partial x^m} + \frac{\partial g_{mq}}{\partial x^p} - \frac{\partial g_{mq}}{\partial x^q} \right)$$

$$+ \frac{1}{2} \left(\frac{\partial g_{qp}}{\partial x^m} + \frac{\partial g_{mp}}{\partial x^q} - \frac{\partial g_{qm}}{\partial x^p} \right) = \frac{\partial g_{pq}}{\partial x^m}.$$

(c) $\frac{\partial}{\partial x^m} (g^{jk} g_{ij}) = \frac{\partial}{\partial x^m} (\delta^k_i) = 0.$ Then

$$g^{jk} \frac{\partial g_{ij}}{\partial x^m} + \frac{\partial g^{jk}}{\partial x^m} g_{ij} = 0 \quad \text{or} \quad g_{ij} \frac{\partial g^{jk}}{\partial x^m} = - g^{jk} \frac{\partial g_{ij}}{\partial x^m}.$$

Multiplying by g^{ir}, $g^{ir} g_{ij} \frac{\partial g^{jk}}{\partial x^m} = - g^{ir} g^{jk} \frac{\partial g_{ij}}{\partial x^m}$

i.e. $$\delta^r_j \frac{\partial g^{jk}}{\partial x^m} = - g^{ir} g^{jk} ([im,j] + [jm,i])$$

or $$\frac{\partial g^{rk}}{\partial x^m} = - g^{ir} \left\{ {k \atop im} \right\} - g^{jk} \left\{ {r \atop jm} \right\}$$

and the result follows on replacing r, k, i, j by :

p, q, n, n respectively.

V. Determine the Christoffel symbols of the second kind in cylindrical coordinates.

In cylindrical coordinates, $x^1 = \rho$, $x^2 = \phi$, $x^3 = z$, we have $g_1 = 1$, $g_{22} = \rho^2$, $g_{33} = 1$.
The only non-zero Christoffel symbols of the second kind can occur where p = 2, these are

$$\left\{ {1 \atop 22} \right\} = -\frac{1}{2g_{11}} \frac{\partial g_{22}}{\partial x^1} = -\frac{1}{2} \frac{\partial}{\partial \rho} (\partial^2) = -\rho ,$$

$$\left\{ {2 \atop 21} \right\} = \left\{ {2 \atop 12} \right\} = \frac{1}{2g_{22}} \frac{\partial g_{22}}{\partial x^1} = \frac{1}{2\rho^2} \frac{\partial}{\partial \rho} (\rho^2) = \frac{1}{\rho} .$$

24. Exercises on tenors

1. Write each of the following using the summation convention

   (a) $(x^1)^2 + (x^2)^2 + \ldots + (x^N)^2$,

   (b) $d\,{}^2 = g_{11}(dx^1)^2 + g_{22}(dx^2)^2 + g_{33}(dx^3)^2$,

   (c) $\sum_{p=1}^{3} \sum_{q=1}^{3} g_{pq}\, dx^p\, dx^q$.

2. Write the terms in each of the following indicated sums

   (a) $A_{pq} \; A^{qr}$

   (b) $\bar{g}_{rs} = g_{jk} \frac{\partial x^j}{\partial \bar{x}^r} \frac{\partial x^k}{\partial \bar{x}^s}$, $N = 3$.

3. Show that $\frac{\partial A_p}{\partial x^q}$ is not a tensor even though A_p is a covariant tensor of rank one.

4. Show that the velocity of a fluid at any point is a contravariant tensor of rank one.

5. If $\bar{A}^p = \frac{\partial \bar{x}^p}{\partial x^q} A^q$, prove that $A^q = \frac{\partial x^q}{\partial \bar{x}^p} \bar{A}^p$.

6. Prove that δ^p_q is a mixed tensor of the second rank.

7? If a tensor A^{pqr}_{st} is symmetric (skew-symmetric) with respect to indices p and q in one coordinate system, show that it remains symmetric (skew-symmetric) with respect to p and q in any coordinate system.

8. Show that every tensor can be expressed as the sum of two tensors, one of which is symmetric and the other skew-symmetric in a pair of covariant or contravariant indices.

9. If A^{pq}_r and B^s_t are tensors, prove that $C^{pqs}_{rt} = A^{pq}_r B^s_t$ is also a tensor.

10. Let A^{pq}_{rst} be a tensor. (a) Choose p = t and show that A^{pq}_{rsp}, where the summation convention is employed, is a tensor. What is its rank? (b) Choose p = t and q = s and show similarly that A^{pq}_{rqp} is a tensor. What is its rank?

11. Prove that the contraction of the tensor A^p_q is a scalar or invariant.

12. A quantity A (p, q, r) is such that in the coordinate system x^i, $A(p, q, r)\, B_r^{qs} = C_p^s$ where B_r^{qs} is an arbitrary tensor and C_p^s is a tensor. Prove that A(p, q, r) is a tensor.

13. Show that the contraction of the outer product of the tensors A^p and B_q is an invariant.

14. Determine the conjugate metric tensor in

    (a) Cylindrical and (b) Spherical coordinates.

15. Prove that g^{jk} is a symmetric contravariant tensor of rank two.

16. (a) show that $L^2 = g_{pq} A^p A^q$ is an invariant.

    (b) Show that $L^2 = g^{pq} A_p A_q$.

17. Prove that for an orthogonal coordinate system $g_{11} = \frac{1}{g^{11}}$, $g_{22} = \frac{1}{g^{22}}$, $g_{33} = \frac{1}{g^{33}}$.

18. Prove (a) $[pq,r] = [pq,r]$, (b) $\left\{ {p \atop pq} \right\} = \frac{\partial}{\partial x^q} \log \sqrt{g}$.

19. Derive the transformation laws for the Christoffel symbols of (a) the first kind, (b) the second kind.

20. Evaluate the Christoffel symbols of (a) the first kind, (b) the second kind, for spaces where $g_{pq} = 0$ if $p \neq q$.

21. Determine the Christoffel symbols of the second kind in (a) rectangular and (b) spherical coordinates.

22. If A_p and A^q are tensors show that.

(a) $A_{p,q} \equiv \frac{\partial A_p}{\partial x^q} - \left\{ \begin{matrix} s \\ pq \end{matrix} \right\} A_s$, and

(b) $A^p_{,q} \equiv \frac{\partial A^p}{\partial x^q} + \left\{ \begin{matrix} p \\ qs \end{matrix} \right\} A^s$ are tensors.

23. Prove that the covariant derivatives of g_{jk}, g^{jk} and δ^j_k are zero.

CHAPTER 5: THE SPECIAL THEORY OF RELATIVITY

1. Breaking up the Newtonian

We are now faced with the task of breaking up the pattern of Newtonian mechanics to make way for the new pattern of relativity, which we owe to Einstein.

To show how fundamental the change is, we shall describe an imaginary experiment, putting into opposition the predictions that would be made by a follower of Newton on one hand and a follower of Einstein on the other. Two clocks stand side by side at a place P. They are of the very finest construction and identical with each other. Their reading are the same, and they continue to run in perfect as long as they stand side by side at P, one clock 'is left at P; the other is put in an airplane and flown with great speed on a long flight, being finally brought back to P and set up beside the clock that has stood there unmoved.

Will there then be any difference between the readings of the two clocks?

The practical physicist before answering make; inquiries as to the way in which the clock was treated on the flight-whether it was knocked about, whether it was subjected to extremes of heat and cold and so on. Let us suppose that the greatest care has been taken, so that effects due to these accidental causes may be ruled out of consideration. Then the answers are as follows:

Newtonian Theory: The clocks will show the same reading.
Relativity Theory: The readings will not be the same.

The clock that has been on the flight will be slow in comparison with the clock that has stayed at home. The follower of Newton reasons along these lines. A perfect clock registers the time. A flight in an airplane does not alter this fact, provided that proper precautions are taken. Since after the flight each clock registers the time, they must agree.

We cannot yet give the reasoning of the relativist; that will come later. For the present, we must be satisfied with the words with which the relativist would bring his attack on the argument of the Newtonian: There is no such thing as the time in any absolute sense. It would be impossible to decide between the two predictions by carrying out the experiment we have described. The relativist would predict a difference between the two readings far too small to detect. The airplane would have to fly with a speed comparable to that of light before the effect would be noticeable. But it is the principle that is important. Other experiments can be carried out in which the predictions of the two theories are different and the difference is large enough to measure; in every case the relativist prediction proves correct. There can be no doubt that the theory of relativity gives us a mathematical model closer to nature than the Newtonian model. We must therefore pay attention to the words: There is no such thing as the time, in any absolute sense. Once that point is conceded, the basis of the Newtonian pattern is broken and the way is open for relativity.

2. The Ingredients of relativity

The theory of relativity is divided into two parts:
(i) The special theory,
(ii) The general theory.

The special theory deals with phenomena in which gravitational attraction plays no part, while the general theory might be called "Einstein's theory of gravitation". We shall be concerned with the special theory.

First we introduce a particle, next we introduce a frame of reference and an observer in it. The observer has a measuring rod with which he can measure the distances between the particles which form his frame of reference. If the distances between these particles remain constant, the observer declares that his frame of reference is a rigid body. Now we provide the observer with a clock. As we have indicated above, the transporting of a clock is an operation which may lead to curious consequences. We shall therefore not expect the observer to carry his clock about but shall provide him with a great number of clocks, all of identical construction. These will be distributed throughout his frame of reference and kept fixed in it.

We must not overlook the fact that the synchronization of these clocks raises an important and difficult question. If there is no synchronization, the observer will note some strange things as he walks among his clocks. For example, he may start at 2:15 (by the local clock), walk a mile, and find that the time is 2:10 (by the local clock). Under such circumstances, in ordinary life one would put a clock in his pocket and walk around, setting each local clock as he passed to agree with the clock in his pocket.

But if our observer does this he finds the following strange result. The clock which he synchronizes in walking out from his base no longer agrees with the clock in his pocket when he is walking back.

This is the same phenomenon as that described earlier in the case of the clock and the airplane, and the reason for it will be made clear later. The effects are so small as to be negligible in ordinary life; but our observer is expected to be **mathematically accurate**. Familiar ideas are turned upside down. Why does the observer not, simply set all the clocks to show the correct time? The answer is: there is no such thing as the correct time.

We introduce the idea of an **event**-some thing happening suddenly at a point. We carry this idea over into relativity, where we shall make extensive use of it. Even though his clocks are not yet synchronized, the observer is prepared to describe any event by assigning four coordinates to it. Of these coordinates, three are spatial (x, y, **z**), and the fourth (t) is given by the local clock, i.e., the clock situated at the point where the event occurs.

3. Galilean Frames of Reference

We have already seen in Newtonian mechanics the importance of making a proper choice of frame of reference. The laws of Newtonian mechanics take their simplest form only in certain special frames, which we called Newtonian. Similarly, in relativity there are frames of reference which are particularly convenient to use. These are called Galilean. frames of reference, They correspond in nature to rigid bodies situated in remote space, far from attracting matter and without rotation relative to the stars as a whole. We shall now make the following hypothesis regarding a Galilean frame of reference:

I. A Galilean frame of reference is a rigid body, isotropic with respect to mechanical and optical experiments.

To explain this, we note that isotropic means, the same in all directions. The neighborhood of the earth is not isotropic. If we drop a stone, it falls in a definite direction and the earth's rotation defines a direction which we can detect by means of a gyrocompass. However, in applying the theory of relativity, we may often regard the earth as a Galilean frame, for its gravitational attraction may be small compared with other forces involved and its rotation may be of no importance. The assumption - that a rigid body in remote space is isotropic is at least plausible, To assert that it was not isotropic would at once raise the question:

Why should any one direction be privileged above another?

The hypothesis refers to mechanical and optical experiments. We must provide the observer with apparatus to perform these. We shall therefore give him mechanisms by which he can exert forces, and lamps and mirrors by which he can send out flashes of light and reflect them. In Newtonian mechanics, we had no occasion to refer to light. Optics appeared to be a separate subject. In relativity, on other hand, we have to discuss optics and mechanics together.

4. The synchronization of Clocks

To reach the most interesting deductions quickly, we shall outline some steps in the development without proof. Thus, we shall only sketch the method of synchronization of clocks in a Galilean frame of reference. The synchronization is done by means of light signals. Taking the clock at the origin O as master clock, the observer sends out flashes of light to the other clocks, from which they are reflected by mirrors back to O. Let t1 and t2 be the times (as given by the clock at O) at which a flash leaves O and returns to it after reflection at a point A. In ordinary life, we should reason in this way: If v is the velocity of light and r the distance OA, the light would take a time r/v to go and a time r/v to return. Thus t2- t1=2r/v, and the time of arrival at A is:

$$t = t_1 + \frac{r}{v} = t_1 + \frac{1}{2}(t_2 - t_1) = \frac{1}{2}(t_1 + t_2)$$

But we cannot use this argument, because velocity is a derived concept, depending on the measurement of both **distance and time**. We should be arguing in a circle if we used velocity to define time. We shall merely adopt the definition of synchronization that the clock at A is synchronized when it is set to read (t1+t2)/2 at the instant when the flash strikes it. By this rule, all the clocks may be synchronized with the clock at O.

5. The Lorentez Transformation

5.1 The Principle of Equivalence

Let us suppose that we can shoot a rocket right out of the solar system. In this rocket we place an observer. When the rocket has passed far beyond the solar system, it forms a Galilean frame of reference. The observer has lost the sense of motion he had when rushing past the planets. He sees around him nothing but stars, and they are so far away that they appear fixed.

A second identical rocket is shot out with a greater speed and on such a track that it overtakes the first. In it there is also an observer. Now we have two Galilean frames of reference.

Imagine that the two observers leave their rocket and travel independently in space. One of them comes upon one of the rockets. How is he to tell whether it is the rocket he occupied before or the other one? To answer this question, he is allowed to perform any mechanical or optical experiments he chooses. The same question in a different form occurred to the physicist Michelson. What he asked might be put thus: Is it possible to tell the season of the year (i.e., the position of the earth in its orbit round the sun) by means of optical experiments performed on a clouded earth? The earth at the two seasons corresponds to the two rockets. It was fully expected that the season could be determined in this way, for it was then believed that light was propagated in an "ether", and the difference between the velocities of the earth through the ether at the two seasons should be a measurable quantity.

The question was yet to experimental test by Michelson and later by Michelson and Morley. The expected result was not obtained. As far as this experiment was concerned, the two seasons were indistinguishable. Generalizing from the negative result of the Michelson— Morley experiment, we make the following sweeping hypothesis:

II. Principle of equivalence: Two Galilean frames of reference are completely equivalent for all physical experiments. This gives the answer to the question raised earlier. The observer is not able to tell which rocket he has found. No experiment he can perform will tell him which it is - they are indistinguishable, like identical twins. To the hypothesis already made we add another:

III. Any two Galilean frames of reference have, relative to one another, a uniform velocity of translation. The relative velocity is less than the velocity of light.

5.2. The Lorentz Transformation

Let S and S' be two Galilean frames of reference. Consider any event, observed by both observers. To this event, S attaches coordinates (x, y, z, t), and S' attaches coordinates (x', y', z', t'). In doing this, each observer uses his own measuring rod and clocks. The event determines the coordinates, and conversely the coordinates determine the event. Thus, considering all possible events, four numbers (x, y, z, t) determine an event, and that in turn determines the four numbers (x', y', z', t'). The last four numbers are therefore functions of the first four and we express this by writing:

$x'=f(x,y,z,t),\ y'=g(x,y,z,t),\ z'=h(x,y,z,t),\ t'= l(x, y, z,t)$------------*(1)*

Such relations, connecting the coordinates of two Galilean observers, constitute a Lorentz transformation. We have now to investigate the forms of these functions. We shall not, however, suppose that the axes oxyz and o'x'y'z' are given arbitrary directions in the respective frames. We shall suppose them so chosen that ox and o'x' be on a common line when viewed by either observer, this line being parallel to the relative velocity of either frame with respect to the other.

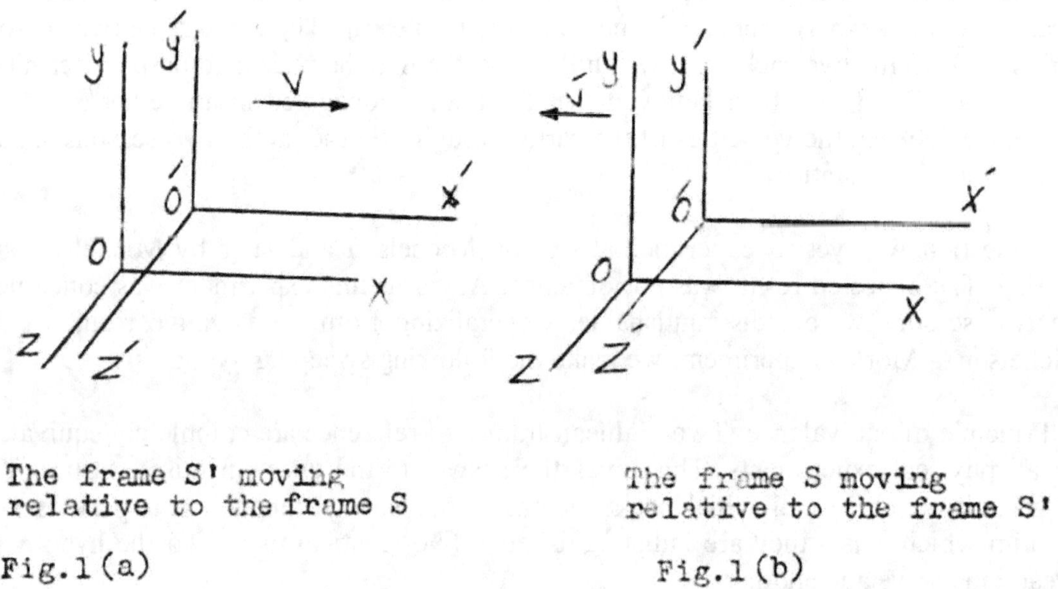

We shall consider the Lorentz transformation only for events occurring on this common line. Thus y = z = y' = z' = 0, and the transformation is of the form:

$x' = f(x,t)$, $t' = l\,(x,t)$ ------------(2)

We must carefully avoid the idea that there is any absolute frame from which S and S' may be viewed. We must look at things either as they appear to S or as they appear to S'. First, S sees the particles of his own frame. They are fixed as far as he is concerned, and through them there pass the particles of the frames S'. These particles all move parallel to ox with a constant speed v, the speed of S' relative to S. Similarly, to S' the particles of his frame appear fixed, and the particles of S pass with a speed v', directed in the negative sense of o'x'. To do justice to both observers, it is best to draw two diagrams, as in Figs. 1(a) and 1(b). In Fig.1(a) we take the view of S and regard oxyz as stationary; in Fig.1(b) we take the view of S' and regard o'x'y'z as stationary. The two observers now fix their attention on a flash of light traveling along the common line ox, o'x'. It will add to the complexity of our work if we assume that the units of space and time used by S and. S' are completely independent. We shall therefore suppose that they have been supplied with measuring rods and clocks. Then, by virtue of the principle of equivalence, the speed of light has a common value in the two frames. This common value we shall denote by c.

As S observes the flash, he records the time t at which the flash reaches the position x. Since the speed of light is c, therefore, x is a function of t satisfying:

$$\left(\frac{dx}{dt}\right)^2 = c^2.$$

But similarly, for the observations of S',

$$\left(\frac{dx'}{dt'}\right)^2 = c^2.$$

Thus, for the sequence of events given by the passage of the flash we have the two equations:

$$dt^2 - \frac{dx^2}{c^2} = 0\ , \qquad dt'^2 - \frac{dx'^2}{c^2} = 0\ .$$

Now a motion satisfying either of these equations represents the passage of a flash of light and therefore must satisfy the other equation. Thus, if one of the equations is satisfied, so is the other, and therefore we have the identity:

$$dt'^2 - \frac{dx'^2}{c^2} \equiv K\left(dt^2 - \frac{dx^2}{c^2}\right), \tag{3}$$

where K is some unknown factor. But by the principle of equivalence we must also have

$$dt^2 - \frac{dx^2}{c^2} \equiv K\left(dt'^2 - \frac{dx'^2}{c^2}\right). \tag{4}$$

Comparing these two identities, we see that $K^2 = 1$, and so $K = \pm 1$. To see which value to take, we follow the particle O', fixed in S'. Then $dx' = 0$, and so, by (3),

$$\left(\frac{dt'}{dt}\right)^2 = K\left[1 - \frac{1}{c^2}\left(\frac{dx}{dt}\right)^2\right].$$

But, by the hypothesis III, $\left(\frac{dx}{dt}\right)^2 < c^2$; hence $K = +1$. Accordingly

$$dt'^2 - \frac{dx'^2}{c^2} \equiv dt^2 - \frac{dx^2}{c^2}. \qquad (5)$$

The Lorentz transformation must be such that this identity holds. We shall assume that the transformation is linear and that the zeros of time are so that t t' = 0, when 0' is passing through O. Thus we write in place of (2),

$$x' = \alpha x + \beta t \qquad t' = \alpha' x + \beta' t, \qquad (6)$$

where $\alpha, \beta, \alpha', \beta'$ are constants so connected that the identity (5) is satisfied. We have,

$$dx' = \alpha\, dx + \beta\, dt, \quad dt' = \alpha'\, dx + \beta' dt,$$

$$dt'^2 - \frac{dx'^2}{c^2} = (\alpha'\, dx + \beta' dt)^2 - (\alpha dx + \beta\, dt)^2/c^2$$

$$= dt^2 - \frac{dx^2}{c^2} \qquad (7)$$

and so, equating the coefficients of dx^2, dt^2, and $dxdt$,

$$\alpha^2 - c^2\alpha'^2 = 1, \quad \beta^2 - c^2\beta'^2 = -c^2,$$
$$\alpha\beta - c^2\alpha'\beta' = 0. \qquad (8)$$

Let us define ϕ, ϕ' by the equations.

$$\sinh \phi = C\alpha', \quad \sinh \phi' = \beta/C \,. \tag{9}$$

Then, by the first two equations in (8), we have

$$\alpha = \cosh \phi, \quad \beta' = \cosh \phi',$$

and the last of (8) gives:

$$\sinh(\phi' - \phi) = 0$$

Thus $\phi' = \phi$, and the transformation (6) is

$$x' = x \cosh \phi + ct \sinh \phi,$$

$$ct' = x \sinh \phi + ct \cosh \phi. \tag{10}$$

It is easy to verify directly that this transformation satisfies (5) for any constant ϕ.

To identify ϕ, we take the viewpoint of S and follow the particle O'. For O' we have $x' = 0$, $\frac{dx}{dt} = V$, and so differentiation of the first of (10) gives:

$$\tanh \phi = -\frac{V}{C} \tag{11}$$

Hence

$$\cosh \phi = \frac{1}{\sqrt{1 - \frac{V^2}{C^2}}} \quad , \ \sinh \phi = \frac{-\frac{V}{C}}{\sqrt{1 - \frac{V^2}{C^2}}} \quad , \tag{12}$$

and so we have the Lorentz transformation ,

$$x' = \gamma(x - Vt), \quad t' = \gamma\left(t - \frac{Vx}{c^2}\right), \quad \gamma = \frac{1}{\sqrt{1 - \frac{V^2}{c^2}}}. \qquad (13)$$

Solving for x, t, we get:

$$x = \gamma(x' + Vt'), \quad t = \gamma\left(t' + \frac{Vx'}{c^2}\right). \qquad (14)$$

If we now take the viewpoint of S' and follow O, we have x = 0, and $\frac{dx'}{dt'} = -V'$,

where V' is the speed of S relative to S'. But when we put x=0 in the first of (14) and differentiate we obtain $\frac{dx'}{dt'} = -V$.

Hence V' = V, as indeed we might have anticipated from the principle of equivalence. Now we have the explanation why the theory of relativity did not force itself on the attention of mankind long ago. High-speed particles (electrons protons, etc.) were not observed until comparatively recent times, and physicists dealt only with relative velocities very small indeed compared with the velocity of light. If v/c is small, then the Lorentz transformation (13) tends to

$$x' = x - vt, \qquad t' = t \text{ ---------- } (15)$$

as in Newtonian mechanics.

5.3. Immediate Consequences of the Lorentz Transformation

To a person accustomed to thinking in the Newtonian way; some of the predictions of the theory of relativity are startling.

Outstanding among these are the contraction of a moving body and the slowing down of a moving block. These apparently curious facts are consequences of the Lorentz transformation (13). First, let us consider a measuring rod which S' lays down along his axis o'x'. To him it is a fixed measuring rod. If A, B are the ends, the history of A is a sequence of events for which x = (x 2) A' a constant, and the history of B is a sequence of events for which $x' = (x')_A$, a constant, and the history of B is a sequence of events for which $x' = (x')_B$, also a constant; the length of the rod is:

$$L' = (x')_B - (x')_A \cdot \tag{16}$$

Viewed by S, the rod is not fixed. An instantaneous picture, taken by S at time t, shows A at $(x)_A$, say, and B at $(x)_B$. If asked what is the apparent length of the rod, S says that it is:

$$L = (x)_B - (x)_A \cdot \tag{17}$$

Now, by the first equation of (13),

$$\left.\begin{aligned} (x')_B &= \gamma\left[(x)_B - Vt\right] \\ (x')_A &= \gamma\left[(x)_A - Vt\right] \end{aligned}\right\} \tag{18}$$

and subtraction gives, in view of (16) and (17),

$$L' = \gamma L,$$

$$\text{or} \quad L = \frac{L'}{\gamma} = L'\sqrt{1 - \frac{V^2}{c^2}} < L'. \tag{19}$$

Thus L is less than L'; the rod appears to S to be contracted in the ratio: $\sqrt{1 - \frac{V^2}{c^2}} : 1.$

We might expect that, if S' viewed a rod fixed in S, he would see an expansion instead of a contraction. But if we carry out the calculation, now using (14) instead of (13), we find the same contraction again. Each observer considers that the measuring rod of the other is contracted.

Now let us consider a clock carried along in S'. Let $(t')_A$, $(t')_B$ be two readings of the clock. These are two events, both with the same x', and with $t' = (t')_A$, $t' = (t')_B$, respectively Viewed by S, the clock is moving, let the times of the two events be $(t)_A$, $(t)_B$ respectively, as measured in his time system. From the second equation of (14), we have,

$$\left.\begin{aligned} (t)_B &= \gamma \left[(t')_B + \frac{Vx'}{c^2}\right], \\ (t)_A &= \gamma \left[(t')_A + \frac{Vx'}{c^2}\right] \end{aligned}\right\} \qquad (20)$$

By sub traction,

$$(t)_B - (t)_A = \gamma \left[(t')_B - (t')_A\right],$$

or

$$T = \gamma T' = \frac{T'}{\sqrt{1 - \frac{V^2}{c^2}}} > T',$$

where T is the time interval recorded by S and T' the time interval recorded by S'. Since T'<T, the clock carried by S' appears to S to be running slow. Just as in the case of contraction of length, this result works both ways. Each observer considers the clock of the other to be running slow.

6. Space-time

It does not seem possible at first sight to do justice simultaneously to each of two Galilean observers, for it appears necessary to take the point of view of either the one or the other. This difficulty is overcome by using a space-tine diagram.

Consider first one Galilean observer S. Draw oblique cartesian axes and label them Ωx, Ωt (Fig. 2). (We use Ω instead of 0 for origin to avoid confu e ion with origin of the observer's axes). Any event which happens on the axis of ox in the observer's frame will have attached to it values of x and t. It can be represented by a point in Fig. 2, which is our space-time diagram. The history of a particle moving along Ox will appear as a curve C in the space-tine diagram. If it moves with uniform velocity, dx/dt is a constant, and the curve becomes a straight line Cl

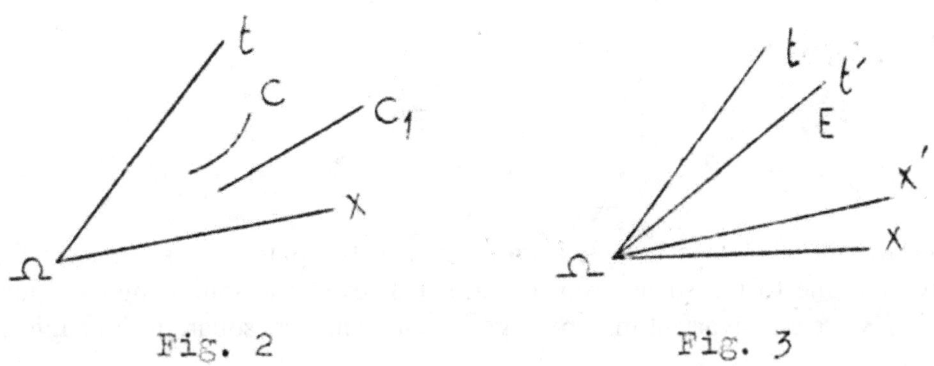

Fig. 2 Fig. 3

L Fig. 2 Fig. 3 Consider now a second Galilean observer . Instead of making a new space-time diagram for him, we superimpose his

diagram on that of S. But we use new axes Ω x't' (Fig. 3) so related to Ω xt that the geometrical law of transformation of

coordinates in the plane is precisely the Lorentz transformation (13). Now any even (has of course two pairs of labels (x,t), (x',t') but since these are connected by the Lorentz transformation, E appears as a single point in the space-time diagram. In ordinary methods of geometry, we have two sets of rectangular axes oxy, o'x'y' in a plane, then the square of the distance between adjacent points is:

$$(dx)^2 + (dy)^2 = (dx')^2 + (dy')^2 \qquad (22)$$

In fact, the expression $(dx)^2 + (dy)^2$ is an invariant. But in the space-time diagram,

$$(dt)^2 + (dx)^2 \neq (dt')^2 + (dx')^2.$$

The invariant quantity is by (5),

$$(dt)^2 - (\frac{dx}{c})^2 = (dt')^2 - (\frac{dx'}{c})^2 \qquad (23)$$

In geometry, we denote the invariant (22) by $(ds)^2$ and

call dS thc distancc bctwccn two adjaccnt points in thc planc. This suggcsts that wc should give a name to the square root of (a3). However, the minus sign introduces a complication, since the invariant may be negative and hence its square root imaginary. So we put:

$$\epsilon\ (dS)^2 = (dt)^2 - (\frac{dx}{c})^2 \qquad (24)$$

where $\epsilon = +1$ or -1, according as the expression on the right is positive or negative. We call dS the separation between the events (x,t) and $(x + dx,\ t + dt)$. In the case of two events which are not adjacent, we define the separation as $S = \int dS$, taken along the straight line joining the points in the space-time diagram which correspond to them. Since $\frac{dx}{dt}$ is constant along a straight line, we easily find,

$$\epsilon\ S^2 = (t_1 - t_2)^2 - (x_1 - x_2)^2/c^2, \qquad (25)$$

where the two events in question are (xl, tl) and (x2, t2). We note that if two events have the same x, the separation is simply the difference between the t's; if they have the same t.

The separation is the difference between the x.'s divided by c. The lines in the space-time diagram satisfying one or other of the equations, $dt = dx/c$, $dt' = dx'/c$ --------- (26)

are called null lines, because the separation between any two points on such a line is zero.

We shall postpone the discussion of motion under a force to the next section, but now accept the law that a free particle travels in a straight line with constant speed, just as in Newtonian mechanics. This means that a free particle traveling along ox moves in accordance with:

$$dx/dt = u$$

where u is a constant. Its history appears in the space-time diagram as a straight line with equation,

$$t = -x/u = \text{constant.} \quad \text{------------------(27)}$$

We shall now show how the contraction of a moving rod and the slowing down of a moving clock appear in the diagram.

In Fig. 4 the lines a, b represent the histories of the ends of a measuring rod lying on Ox' and fixed in S'. These are drawn parallel to $\Omega t'$ because the x' of each end of the rod is a constant. The length L' (judged by S') is proportional to the separation A'B'. To get an instantaneous picture from the viewpoint of S, we draw a line parallel to Ωx (i.e. with t constant), cutting a, b at A, B, respectively. Then the length L (as judged by S) is proportional to the separation AB. That $L \neq L'$ is evident from the diagram. The line a in Fig. 4 may also be regarded as the history of a clock

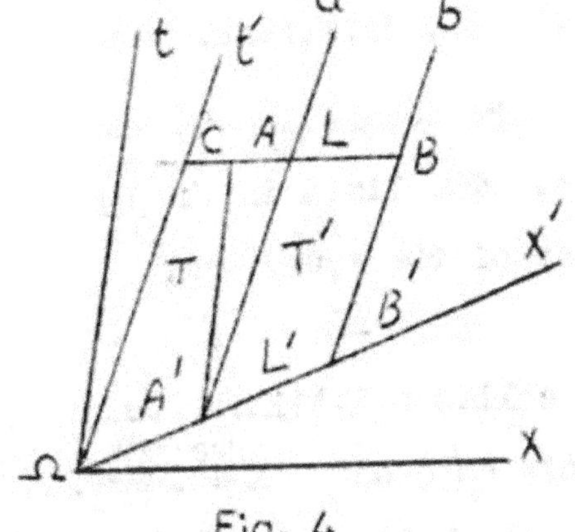

Fig. 4

fixed in S'. In its history, A', A are two events, and the time interval T' between them (as judged by S') is the separation A'A. As judged by S, however, the interval between these events is T, the increase in t in passing from A' to A; it is, in fact, the separation of A'C, where A'C is drawn parallel to Ωt. It is evident that T'/T.

7. Kinematics and Dynamics of A Particle

Composition of velocities: Let P, Q be two particles traveling with uniform vclocitics u1, u2 along the axis ox of a Galilean frame of reference S. What is the relative velocity of the two particles?

In Newtonian mechanics, we should answer: u2-u1. In relativity, we say that this is only the difference between the velocities. The velocity of Q relative to P is the velocity of Q as estimated by a Galilean observer S' traveling along with P, i.e. using a Galilean frame of reference in which P is fixed.

Between S and S' we have the Lorentz transformation,

$$\left.\begin{aligned} x' &= \gamma_1(x - u_1 t), \qquad t' = \gamma_1\left(t - \frac{u_1 x}{c^2}\right), \\ \gamma_1 &= \frac{1}{\sqrt{1 - u_1^2/c^2}}. \end{aligned}\right\} \qquad (1)$$

Consider now the motion of Q; for it, dx/dt = u2 and its velocity as estimated .by S' is.

$$u' = \frac{dx'}{dt'} = \frac{dx - u_1 dt}{dt - u_1 \frac{dx}{c^2}} = \frac{u_2 - u_1}{1 - u_1 u_2/c^2}. \qquad (2)$$

This is the law which replaces the Newtonian law,

$u' = u_2 — u_1$. ------------- (3)

We note that, if ul and u2 are small compared with c; (2) differs very little from (3). If we solve (2) for u 2 and then make a change in notation, we get the relativistic law of composition of velocities: If a particle moves with velocity u1 in S', and S' has a velocity u2 relative to S I then the velocity of the particle relative to S is:

$$\frac{u_1 + u_2}{1 + u_1 u_2 / c^2} . \qquad (4)$$

8. Proper time

Let S be a Galilean frame of reference, and let P has a particle traveling along the axis ox with uniform velocity u. It may be regarded as a particle of a second Galilean frame S'. Consider two a adjacent events in the history of P; the time interval dtt between them is by equation (13) [last part].

$$dt' = \frac{dt - u\frac{dx}{c^2}}{\sqrt{1 - \frac{u^2}{c^2}}}$$

But $\frac{dx}{dt} = u$, and so

$$dt' = dt\sqrt{1 - \frac{u^2}{c^2}} = \sqrt{dt^2 - \frac{dx^2}{c^2}} = ds ,$$

where ds is the separation between the two events. Thus the separation equals the time interval, as measured by a clock carried with the particle.

We now think of a particle traveling with accelerated motion. Its history appears in the space-time diagram as a curve, How does a clock behave if carried with the accelerated particle?

Our previous hypotheses do not tell us; we must make a new assumption. This assumption is as follows:

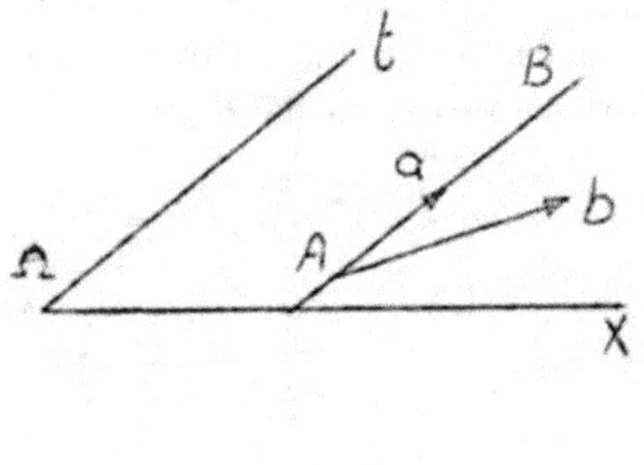

Fig. 5

The time interval between two adjacent events in the history of an accelerated clock is given by the separation between these events.

This separation is called the interval of proper time between the events. Thus the element of proper time for a particle moving along ox with speed u is:

$$dS = \frac{dt}{\gamma u}$$

$$\text{where} \quad \gamma_u = \frac{1}{\sqrt{1 - \frac{u^2}{c^2}}} \,. \tag{6}$$

Now we can give the explanation of the curious prediction made early in the section regarding the behavior of a clock taken on a flight. Figure 5 is a space—time diagram; a is the clock that stayed at home. At the event A(t = t1) the other clock left and described the space—time curve b, returning at the event B(t = t 2). Suppose that both clocks reads zero at A. Then at B the clock that stayed at home reads (since dx = 0 throughout its history).

$$T = \int_A^B dS \qquad \text{(along a)}$$

$$= \int_{t_1}^{t_2} dt = t_2 - t_1 \qquad (7)$$

The clock that flew reads:

$$T' = \int_A^B dS \qquad \text{(along b)}$$

$$= \int_{t_1}^{t_2} \frac{dt}{\gamma_u}$$

$$< t_2 - t_1$$

Thus T' < T, which proves the validity of the Prediction.

9. Equations of motion in absolute form

It has been remarked that gravitation lies outside the scope of the special theory of relativity. But there are other forces by means of which accelerated motion may be produced. The difference between the relativistic and the Newtonian predictions are then far too small to measure. The difference becomes appreciable only in the dynamics of atomic particles accelerated by forces of electromagnetic origin. However, the same principle applies throughout, and we may understand it by thinking of any small body under the influence of any force.

We have accepted the hypothesis that all Galilean frames are equivalent. Thus, whatever form of equations of motion one Galilean observer adopts, a similar form must hold for any other Galilean observer. In fact, the equations of motion of a particle must be invariant under the Lorentz transformation.

If we take the Newtonian equation:

$$m \frac{d^2x}{dt^2} = F \qquad (9)$$

and apply the Lorentz transformation (1.3) [last section] to get an equation in x: and tl, we find an equation of quite different form. Thus (9) is not invariant under the Lorentz transformation.

To see what form of equation is suitable, we have to consider space-time vectors. In Fig. (6), we have taken a point A in the space-time diagram and drawn a directed segment $\overrightarrow{\Omega A}$. This is

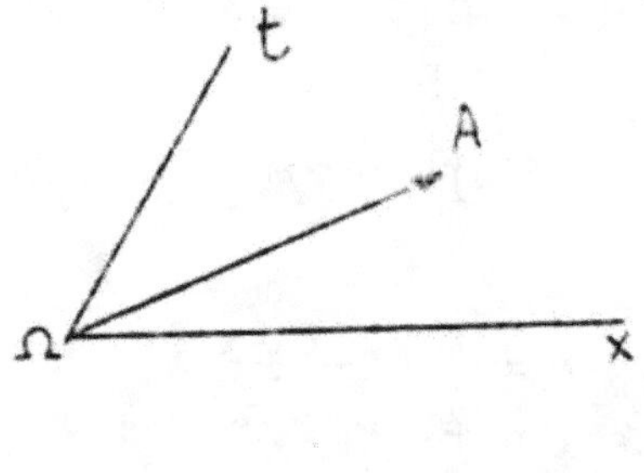

Fig. (6)

a space-time vector; its components in the directions Ωx, Ωt are x,t, the space-time coordinates of A. If we use other axes $\Omega x't'$, the same vector has different components. But between the two sets of components the Lorentz transformation holds.

We define a space-time vector as a pair of quantities (ξ, τ) which transform, when we change axes in space-time just like (x,t), i.e. according to,

$$\left.\begin{aligned} \xi &= \gamma(\xi - v\tau), \quad \tau' = \gamma\left(\tau - \frac{v\xi}{c^2}\right), \\ \gamma &= \frac{1}{\sqrt{1 - \frac{v^2}{c^2}}} \quad . \end{aligned}\right\} \qquad (10)$$

As we have stated our problem is to build equations of motion invariant under a Lorentz transformation. The key to the solution is found in the idea of the space-time vector. We shall form equations in which a space- t vector is equated to a space-time vector. Consider a particle moving along Ox with a general motion. This corresponds to some curve in space time. The proper time, measured from some initial point, may be taken as a parameter, and the equations of the curve written:

$$x = x(S), \quad t = t(S).$$

Since dS is an invariant, the pair of quantities $(\frac{dx}{dS}, \frac{dt}{dS})$ is a space-time vector. We see this by differentiating (13) [last section]. We call $(\frac{dx}{dS}, \frac{dt}{dS})$ the absolute velocity of a particle.

Explicity we have, by ( ),

$$\left.\begin{aligned} \frac{dx}{dS} &= \gamma_u u, \qquad \frac{dt}{dS} = \gamma_u \\ \gamma_u^{-2} &= 1 - \frac{u^2}{c^2}, \end{aligned}\right\} \qquad (11)$$

where $u = \frac{dx}{dt}$, the velocity of the particle relative to the Galilean frame of reference S, corresponding to Ωxt. If the particle has a velocity small compared with that of light, so that $\frac{u}{c}$ is small, the components of the absolute velocity are approximately (u, 1).

Similarly, $\left(\frac{d^2x}{ds^2}, \frac{d^2t}{ds^2}\right)$ is a space-time vector. We call it the absolute acceleration.

We now accept, as satisfactory from the point of view of invariance, under the Lorentz transformation, the following equations of motion:

$$m_o \frac{d^2x}{ds^2} = X \quad , \quad m_o \frac{d^2t}{ds^2} = T \quad , \qquad (12)$$

where m_o is a constant (the proper mass of the particle) and (X, T) is a space-time vector, called the absolute force. In a different Galilean frame of reference S', with velocity V relative to S, these equations read,

$$\left.\begin{aligned} & m_o \frac{d^2x'}{ds^2} = X' \quad , \quad m_o \frac{d^2t'}{ds^2} = T' , \\ \text{where} \quad & X' = \gamma(X - VT), \quad T' = \gamma\left(T - \frac{VX}{c^2}\right) \\ & \gamma = \frac{1}{\sqrt{1 - \frac{V^2}{c^2}}} \end{aligned}\right\} \qquad (13)$$

This may be verified immediately by applying equation (13) (last section) to (12).

10. Equations of motion in relative form

Let us now put the equations of motion (12) into another form in order to show the relation of relativestic to Newtonian mechanics. Since $dS = \frac{dt}{\gamma_u}$, these equations may be written

$$\frac{d}{dt}(m_o \gamma_u u) = \frac{X}{\gamma_u}, \quad \frac{d}{dt}(m_o \gamma_u) = \frac{T}{\gamma_u} \,. \tag{14}$$

Let us define some terms, as follows:

$$\left.\begin{aligned}
&\text{Relative mass} = m = m_o \gamma_u = \frac{m_o}{\sqrt{1 - \frac{u^2}{c^2}}} \,. \\
&\text{Relative momentum} = m u \\
&\text{Relative force} = F = \frac{X}{\gamma_u} = X\sqrt{1 - \frac{u^2}{c^2}} \,. \\
&\text{Relative energy} = E = mc^2 = \frac{m_o c^2}{\sqrt{1 - \frac{u^2}{c^2}}} \,.
\end{aligned}\right\} \tag{15}$$

Then the first of (14) may be written

$$\frac{d}{dt}(mu) = F \,. \tag{16}$$

This is the equation of motion of a particle moving on the x—axis under the influence of a force F. It is of the Newtonian form, but with a remarkable difference. The relative mass of a particle is not a constant, it varies with speed of the particle according to (15). If u/c is small, the variation of m from the value m o is insignificant, but on the other hand, m tends to infinity as the speed of the particle u tends to that of lights. No particle has ever been observed traveling with a speed equal to or greater than that of light.

To discuss the second equation of (14), let us first return to (24) last section. This may be written,

$$\left(\frac{dt}{dS}\right)^2 - \frac{1}{c^2}\left(\frac{dx}{dS}\right)^2 = 1 \,. \qquad (17)$$

Since $\left(\frac{dx}{dt}\right)^2 < c^2$ for a particle. Differentiation gives ,

$$\frac{dt}{dS}\frac{d^2t}{dS^2} - \frac{1}{c^2}\frac{dx}{dS}\frac{d^2x}{dS^2} = 0 \,. \qquad (18)$$

Thus, by (12),

$$T\frac{dt}{dS} - \frac{1}{c^2}X\frac{dx}{dS} = 0 \,, \qquad (19)$$

or

$$c^2T = X u = F u \, \gamma_u \,. \qquad (20)$$

If we multiply the second of (14) by c^2 and substitute from (15) and (20) we get:

$$\frac{dE}{dt} = F u \,, \qquad (21)$$

i.e. rate of change of relative energy = rate of working of relative force .
This has the Newtonian form, but the expression for energy (E) does not at first appear related to the Newtonian kinetic energy. However, if we expand by the binomial theorem, we obtain,

$$E = \frac{m_o c^2}{\sqrt{1 - \frac{u^2}{c^2}}} = m_o c^2\left(1 + \frac{u^2}{2c^2} + \frac{3u^4}{8c^2} + \ldots\right) \,, \qquad (22)$$

and if $\frac{u}{c}$ is small, we have approximately

$$E = m_o c^2 + \frac{1}{2} m_o u^2 \,. \qquad (23)$$

This differs from the Newtonian expression for kinetic energy by the constant $m_o c^2$, which is called the rest energy or proper energy.

This quantity $m_o c^2$ appears of little importance here, because E is differentiated in (21), and so the constant disappears. The equation (21) would still hold if we had adopted the definition,

$$E = \frac{m_o c^2}{\sqrt{1 - \frac{u^2}{c^2}}} - m_o c^2 \qquad (24)$$

for relative energy.

11. Examples on the special theory of relativity

(i) A particle describes the parabolic trajectory, $x = ut$, $y = \frac{1}{2} at^2$ in S. Find its trajectory in the S' system, showing that its acceleration in S' is along O' Y' of magnitude

$$a(1 - \frac{V^2}{c^2}) \;/\; (1 - \frac{uV}{c^2})^2 \; .$$

For S : $x = ut$, $\qquad y = \frac{1}{2} at^2$.

For S': $x' = \gamma[x - Vt]$, $t' = \gamma[t - \frac{Vx}{c^2}]$, $y = \frac{1}{2} at^2$, where, $\gamma = \dfrac{1}{\sqrt{1 - \frac{V^2}{c^2}}}$.

i.e. $x' = \gamma\left[ut - Vt\right]$, $t' = \gamma\left[t - \frac{Vut}{c^2}\right]$,

$y = \frac{1}{2}at^2$

Eliminating t we get,

$$x' = \frac{(u - V)t'}{\left[1 - \frac{Vu}{c^2}\right]} \qquad y' = \frac{at'^2}{2\gamma^2\left[1 - \frac{Vu}{c^2}\right]^2} .$$

Hence

$$\frac{d^2x'}{dt'^2} = 0 , \quad \frac{d^2y'}{dt'^2} = \frac{a}{\gamma^2\left[1 - \frac{Vu}{c^2}\right]^2} = \frac{a\left[1 - \frac{V^2}{c^2}\right]}{\left[1 - \frac{Vu}{c^2}\right]^2} .$$

(ii) In S, a constant force m o g acts on a particle of proper mass m o: in a fixed direction. Write down the relativistic equation of motion and integrate it completely, obtaining:

$$x = \left[(c^2/g)^2 + (ct)^2\right]^{1/2} - \frac{c^2}{g}$$

if $x = 0$ at $t = 0$. Show that this is a hyperbola, and that the classical limit as $c \longrightarrow \infty$ is, $x = \frac{1}{2}gt^2$.

Equation of motion is $\frac{d}{dt}\left(m\frac{dx}{dt}\right) = m_o g$,

where, $m = \dfrac{m_o}{\sqrt{1 - \frac{u^2}{c^2}}}$, $\frac{dx}{dt} = u.$

Hence,

$$\frac{m_o u}{\sqrt{1 - \frac{u^2}{c^2}}} = m_o \int g \; dt = m_o gt .$$

Then $u = \dfrac{gt}{\sqrt{1 + (\frac{gt}{c})^2}}$, solving for u, we get:

$$x = \int \frac{gt\ dt}{\sqrt{1 + (\frac{gt}{c})^2}} = \left[(\frac{c^2}{g})^2 + (ct)^2 \right]^{½} - \frac{c^2}{g} .$$

Squaring, we get

$$(x + \frac{c^2}{g})^2 - c^2t^2 = (\frac{c^2}{g})^2 ,$$

which is a hyperbola of eccentricty $e^2 = 1 + \frac{1}{c^2}$. The classical limit is:

$$x = \frac{c^2}{g} \left[\sqrt{1 + (\frac{gt}{c})^2} - 1\right] = \frac{1}{2} gt^2 - \frac{1}{8} \frac{g^2t^4}{c^2} + \ldots.$$
$$= \frac{1}{2} gt^2$$

as $c \longrightarrow \infty$

(iii) Assuming energy conservation, write down the energy equation for the relativistic oscillator, if the potential energy is given by $\frac{1}{2} m_o \omega^2 x^2$ and the total energy by $\frac{1}{2} m_o \omega^2 a^2$.

Solving this equation for $u = \dot{x}$, and if $x = a \cos \theta$, show that:

$$\dot{\theta} = \frac{\omega \left[1 + \frac{\omega^2 a^2}{4c^2} \sin^2\theta \right]^{½}}{1 + \left[\frac{\omega^2 a^2}{2c^2} \right] \sin^2\theta} .$$

Find ωt as indefinite integral in θ, and by expanding to the first order in $\frac{\omega^2 a^2}{c^2}$, show that

$$\omega t = \left[1 + \frac{3\omega^2 a^2}{16c^2}\right]\theta - \frac{3\omega^2 a^2}{16c^2} \sin\theta\cos\theta,$$

Chosing $t = 0$ when $\theta=0$.

Total energy $= m_o c^2 (\gamma - 1)$ + potential energy; so,

$$\frac{1}{2} m_o \omega^2 a^2 = m_o c^2 (\gamma - 1) + \frac{1}{2} m_o \omega^2 x^2$$

or

$$2c^2 \left[\frac{1}{\sqrt{1 - \frac{u^2}{c^2}}} - 1\right] = \omega^2 (a^2 - x^2)$$

Solving for u,

$$u = \frac{\omega \left[(a^2 - x^2)\left\{1 + \frac{\omega^2}{4c^2}(a^2 - x^2)\right\}\right]^{1/2}}{1 + \frac{\omega^2}{2c^2}(a^2 - x^2)} .$$

If $x = a\cos\theta$,

$$- a\sin\theta\,\dot{\theta} = \frac{\omega}{1 + \frac{\omega^2}{2c^2} a^2 \sin^2\theta} (-1)\left[a^2 \sin^2\theta\right.$$

$$\left.\left(1 + \frac{\omega^2 a^2}{4c^2}\sin^2\theta\right)\right]^{1/2}$$

so $$\dot{\theta} = \frac{\omega}{1 + \frac{\omega^2}{2c^2} a^2 \sin^2\theta} \sqrt{1 + \frac{\omega^2 a^2}{4c^2}\sin^2\theta}\ .$$

Now $\dot{\theta} = \frac{d\theta}{dt}$, so

$$\omega t = \int \frac{1 + \frac{\omega^2 a^2}{2c^2}\sin^2\theta \, d\theta}{\left[1 + \frac{\omega^2 a^2}{4c^2}\sin^2\theta\right]^{1/2}}$$

$$= \int \left(1 + \frac{\omega^2 a^2}{2c^2}\sin^2\theta\right)\left(1 - \frac{\omega^2 a^2}{8c^2}\sin^2\theta\right) d\theta$$

$$= \int \left(1 + \frac{3\omega^2 a^2}{16c^2} - \frac{3\omega^2 a^2}{16c^2}\cos 2\theta\right) d\theta$$

$$= \left(1 + \frac{3\omega^2 a^2}{16c^2}\right)\theta - \frac{3\omega^2 a^2}{16c^2}\sin\theta\cos\theta.$$

---

12. Exercises on the special theory of relativity

1) A light ray is incident at O from a direction making angles 30° with Ox and 60° with Oy, measured in S. If $V = \frac{1}{2}\sqrt{3}\, c$, show that in S' the ray comes from a direction making an angle $\tan^{-1}\left(\frac{1}{4\sqrt{3}}\right)$ with O'x'.

2) In S, events occur at the origin and the point (X,0,0) simultaneously at $t = 0$. The time interval between the events in S' is T. Show that the spatial distance between the events in S' is $\sqrt{X^2 + c^2T^2}$, and find the relative velocity of the two frames in terms of X and T.

3) For suitably chosen axes in two Galilean frames S and S', the complete Lorentz transformation is

$$x' = \gamma(x - Vt), \quad y' = y, \quad z' = z,$$

$$t' = \gamma\left(t - \frac{Vx}{c^2}\right), \quad \gamma = \sqrt{1 - \frac{V^2}{c^2}},$$

where V is the relative velocity of S and S'. A particle, as observed by S', describes a circle $x'^2 + y'^2 = a^2$, $z' = 0$, with constant speed. Show that to S the particle appears to move in an ellipse whose centre moves with velocity V.

4) The history of a moving particle is represented in space-time diagram by the hyperbola.

$$\frac{x^2}{c^2} - t^2 = \frac{1}{k^2} .$$

Show that

$$\frac{d^2x}{ds^2} = k^2x, \quad \frac{d^2t}{ds^2} = k^2t.$$

CHAPTER 6: MATRICES

1. Definition of a matrix

A matrix is a set of mn quantities arranged in a rectangular array of m rows and n columns. In writing down matrices it is usual to enclose this array by square brackets (or round brackets) and to denote the matrix by a single letter, say A, such that

$$A = \begin{bmatrix} a_{11} & a_{12} & \dots & a_{1n} \\ a_{21} & a_{22} & \dots & a_{2n} \\ . & . & & . \\ . & . & & . \\ a_{m1} & a_{m2} & & a_{mn} \end{bmatrix}$$

The individual quantities a_{ik} are called the elements of the matrix, (the first subscript refers to the row **and the** second to the column).

A matrix of m rows and n columns is said to be of order (m x n) and to be square if **m=n.** Unlike determinants, matrices have no algebraic value in fact they must he regarded as operators in the same sense that d/dx is an operator.

2. Algebra of Matrices

a) Addition and subtraction

In order that these operations may be carried out, the matrices concerned must be of the same order. If A and B are two matrices of the same order with elements a_{ik} and b_{ik} respectively then their sum A + B is defined as a matrix C whose elements $c_{ik} = a_{ik} + b_{ik}$ clearly C has the same order as A and B. For example if c_{ik}

$$A = \begin{bmatrix} 1 & -2 & 3 \\ 1 & 0 & 2 \end{bmatrix} \text{ and } B = \begin{bmatrix} 2 & 3 & -1 \\ 0 & 2 & 3 \end{bmatrix}$$

then

$$C = A + B = \begin{bmatrix} 3 & 1 & 2 \\ 1 & 2 & 5 \end{bmatrix}$$

Two matrices of the same order are said to be conformable for addition and subtraction. Subtraction of matrices is defined in a similar way in that the difference of two matrices A and B of the sane order is a matrix D whose elements $d_{ik} = a_{ik} - b_{ik}$ For example using the matrices A and B in the previous example we have

$$D = A - B = \begin{bmatrix} -1 & -5 & 4 \\ 1 & -2 & -1 \end{bmatrix}$$

From the definition it follows that A +B=B+ A
Similarly we have A+ (B + C) = (A + B) + C = A + B + C

So that addition of matrices is both commutative and associative.

b) Equality of matrices

Two matrices A and B with elements *a* and b respectively are only equal if they are of the same order, and if all their corresponding elements are equal (i.e., $a_{ik} = b_{ik}$)

c) Multiplication by a Number (or scalar)

The result of multiplying a matrix A = [a_{ik}] by a number k (real or complex) is defined as a matrix B whose elements b_{ik} are k times the elements of A for example if

The result of multiplying a matrix $A = [a_{ik}]$ by a number k (real or complex) is defined as a matrix B whose elements b_{ik} are k times the elements of A for example if

$$A = \begin{bmatrix} 1 & 2 & 0 \\ 3 & -1 & 2 \end{bmatrix} \quad \text{then } kA = \begin{bmatrix} k & 2k & 0 \\ 3k & -k & 2k \end{bmatrix}$$

From this definition it follows that the distributive law of elementary algebra holds also for matrices in that

$$k(A + B) = kA + kB$$

Furthermore if we define kA = Ak, then multiplication by a number is commutative. It is important to state that k must be a number and not for instance another matrix.

d) Matrix Multiplication

The definition of matrix multiplication A and B can only be multiplied together to form their product AB if the number of columns of A is equal to the number of rows of B. A is then said to be conformable to B for multiplication. Suppose A is a matrix of order (m x p) with elements a_{ik} and B is a matrix of order (p x n) with elements b_{ik}. Then their product AB is a matrix G of order (m x n) with elements c defined by

$$c_{ik} = \sum_{s=1}^{p} a_{is} b_{sk}$$

For example, if A and B are (3x 2) and (2x 2) matrices, respectively, given by

$$A = \begin{bmatrix} a_{11} & a_{12} \\ a_{21} & a_{22} \\ a_{31} & a_{32} \end{bmatrix}, \quad B = \begin{bmatrix} b_{11} & b_{12} \\ b_{21} & b_{22} \end{bmatrix}$$

then the product $C = AB$ is a (3×2) matrix defined as

$$C = \begin{bmatrix} a_{11} b_{11} + a_{12} b_{21} & a_{11} b_{12} + a_{12} b_{22} \\ a_{21} b_{11} + a_{22} b_{21} & a_{21} b_{12} + a_{22} b_{22} \\ a_{31} b_{11} + a_{32} b_{21} & a_{31} b_{12} + a_{32} b_{22} \end{bmatrix}$$

As another example

$$\begin{bmatrix} 2 & 3 & -1 \end{bmatrix} \begin{bmatrix} 1 & 3 \\ 2 & -2 \\ 3 & -1 \end{bmatrix} = \begin{bmatrix} 2+6-3 & 6-6+1 \end{bmatrix} = \begin{bmatrix} 5 & 1 \end{bmatrix}$$

if we take

$$A\ (2 \times 3) = \begin{bmatrix} 3 & 1 & 2 \\ 2 & 1 & 3 \end{bmatrix}, \quad B\ (3 \times 2) = \begin{bmatrix} 1 & 2 \\ 3 & 1 \\ 2 & 3 \end{bmatrix}$$

$$\text{then the product } C\ (2 \times 2) = \begin{bmatrix} 10 & 13 \\ 11 & 14 \end{bmatrix}$$

On the other hand A and B are also conformable for the product B (8x 2), A (2 x 3). This defines a (3 x 3) matrix D given by

$$D = \begin{bmatrix} 7 & 3 & 8 \\ 11 & 4 & 9 \\ 12 & 5 & 13 \end{bmatrix}$$

Clearly AB - BA, since the orders of the matrices representing the two products are different. This non commutative property of matrix multiplication appears even when A and B are such that their two products AB and BA are matrices of the same order. For example if

$$A = \begin{bmatrix} 0 & 1 \\ 1 & 1 \end{bmatrix}, B = \begin{bmatrix} 0 & -1 \\ 1 & 0 \end{bmatrix}$$

then

$$AB = \begin{bmatrix} 1 & 0 \\ 1 & 1 \end{bmatrix} \text{ and } BA = \begin{bmatrix} -1 & -1 \\ 0 & 1 \end{bmatrix}$$

Again $AB \neq BA$ but

$$|AB| = -1, \; |BA| = -1$$

So that $|AB| = |BA|$

However, apart from this non commutative law of multiplication, matrices satisfy the distributive and associative laws of multiplication in that

(i) (A + B) C=-AC + BC
(ii) (AB) C =A (B C)

provided that the products are defined

(i) To show that (A +B) C=AC – BC,

where A and B are of the same order that summation can be performed and the number of columns in A and B are equal to the number of rows in C so that multiplication can be carried out.

Let $A = \begin{bmatrix} x_1 & y_1 & z_1 \\ u_1 & v_1 & w_1 \end{bmatrix}$, $B = \begin{bmatrix} x_2 & y_2 & z_2 \\ u_2 & v_2 & w_2 \end{bmatrix}$

and $C = \begin{bmatrix} a_1 & a_2 \\ b_1 & b_2 \\ c_1 & c_2 \end{bmatrix}$

$$(A+B)\ C = \begin{bmatrix} x_1 + x_2 & y_1 + y_2 & z_1 + z_2 \\ u_1 + u_2 & v_1 + v_2 & w_1 + w_2 \end{bmatrix} \begin{bmatrix} a_1 & a_2 \\ b_1 & b_2 \\ c_1 & c_2 \end{bmatrix}$$

$$\begin{bmatrix} a_1 (x_1 + x_2) + b_1 (y_1 + y_2) + c_1 (z_1 + z_2) & a_2 (x_1 + x_2) + b_2 (y_1 + y_2) + c_2 (z_1 + z_2) \\ a_1 (u_1 + u_2) + b_1 (v_1 + v_2) + c_1 (w_1 + w_2) & a_2 (u_1 + u_2) + b_2 (v_1 + v_2) + c_2 (w_1 + w_2) \end{bmatrix}$$

$$= \begin{bmatrix} a_1 x_1 + b_1 y_1 + c_1 z_1 & a_2 x_1 + b_2 y_1 + c_2 z_1 \\ a_1 u_1 + b_1 v_1 + c_1 w_1 & a_2 u_1 + b_2 v_1 + c_2 w_1 \end{bmatrix} +$$

$$\begin{bmatrix} a_1 x_2 + b_1 y_2 + c_1 z_2 & a_2 x_2 + b_2 y_2 + c_2 z_2 \\ a_1 u_2 + b_1 v_2 + c_1 w_2 & a_2 u_2 + b_2 v_2 + c_2 w_2 \end{bmatrix}$$

$= A\,C + B\,C$

ii) Similarly it can Le proved that (AB) C=A(BC)

provided operations are possible.

Although matrix multiplication is distributive with respect to addition, the failure of the commutative law means that the usual formulae for real numbers concerning special products and factorising are not in general valid for matrices, for example

$$(A+B)^2 = (A+B)(A+B) = A^2 + AB + BA + B^2$$

$$(A+B)(A-B) = A^2 - AB + BA - B^2$$

Example (1)

Determine if possible, the number x. so that the matrices:

$$A = \begin{bmatrix} 3 & 1 \\ x & 3 \end{bmatrix} \text{ and } B = \begin{bmatrix} 2 & 1 \\ 3 & 2 \end{bmatrix}$$

are commutative to multiplication.

These products are equal if and only if x+6=9.

$\therefore$

$x = 3$

For any value of $x \neq 3$, $AB \neq BA$

Example (2)

A manufacturer of furniture wishes to produce 40 tables, 120 chairs, 20 desks and 10 book cases. It requires 8 units of material and 3 man - hours of labor for each table, 4 units of material and 2 man-hours of labor for each chair; 10 units of material and 6 man - hours of labor for each desk; and 12 units of material and 10 man - hours of labor for each book case. The material is £ 1.00 per unit and labor is £ 1.50 per man hour. Find the total cost.

Consider the quantities to be produced as a single row matrix, the amount of material and labor as a (4x2) matrix and the cost as a single column matrix.

$$\text{Total cost} = \begin{bmatrix} 40 & 120 & 20 & 10 \end{bmatrix} \begin{bmatrix} \text{material} & \text{labour} \\ 8 & 3 \\ 4 & 2 \\ 10 & 6 \\ 12 & 10 \end{bmatrix} \begin{bmatrix} 1 \\ 1.5 \end{bmatrix}$$

$$= \begin{bmatrix} \text{material} & \text{labour} \\ 1120 & 580 \end{bmatrix} \begin{bmatrix} 1 \\ 1.5 \end{bmatrix}$$

$$= 1120 \times 1 + 580 \times 1.5 = £\ 1990$$

Example (3)

Express as a single matrix

$$\begin{bmatrix} x & y & z \end{bmatrix} \begin{bmatrix} a & h & g \\ h & b & f \\ g & f & c \end{bmatrix} \begin{bmatrix} x \\ y \\ z \end{bmatrix}$$

This equals to :

$$\begin{bmatrix} ax + hy + gz & hx + by + fz & gx + fy + cz \end{bmatrix} \begin{bmatrix} x \\ y \\ z \end{bmatrix}$$

$$= [\, ax^2 + hxy + gzx + hxy + by^2 + fyz + gxz + fyz + cz^2 \,]$$

$$= [\, ax^2 + by^2 + cz^2 + 2\,fyz + 2\,gzx + 2\,hxy \,]$$

This shows that a homogeneous quadratic form in x, y, z can be expressed as the product of three matrices.

3. Rank of a Matrix

Consider the matrix

$$\begin{bmatrix} a_{11} & a_{12} & a_{13} & \cdots & a_{1n} \\ a_{21} & a_{22} & a_{23} & \cdots & a_{2n} \\ \cdot & \cdot & \cdot & & \cdot \\ a_{m1} & a_{m2} & a_{m3} & \cdots & a_{mn} \end{bmatrix}$$

We say that this matrix is of rank *r* if there exists at least one non zero determinant of order r while all determinants of order higher than r are zero.

The existence of a non - zero determinant of order r implies the existence of at least one non - zero determinant of every order less than r.

The rank of a matrix is of fundamental importance in the study of the linear dependence of equations.

Evidently the rank r ≤ G m and r ≤ n

Example (4)
Determine the rank of the matrix

$$\begin{bmatrix} 1 & 1 & 1 & 1 & 2 \\ 1 & 1 & 1 & 1 & 5 \\ 1 & 1 & 1 & 1 & 6 \\ 1 & 1 & 1 & 1 & 9 \end{bmatrix}$$

It is evident here that all determinants of the fourth order are *zero* since we have in each such determinant two equal columns. For the same reason all determinants of the third order are also zero. We have at least one non zero determinant of order 2, e.g.,

$\begin{vmatrix} 1 & 2 \\ 1 & 5 \end{vmatrix}$. Hence in this case we say that the rank of the matrix is 2

Example (5)
Consider the matrix

$$\begin{bmatrix} a & b & c & d & e \\ p & q & r & s & t \\ a & b & c & d & e \\ p & q & r & s & t \end{bmatrix}$$

Here all determinants of order 4 are zero since we have two equal rows. Similarly all determinants of order 3 are zero. We have at least one non - zero determinant of order 2

e.g. $\begin{vmatrix} d & e \\ s & t \end{vmatrix}$. Hence the matrix is of rank 2. Now suppose

that p, q, r, s, t are proportional to *a,* b, c, d, e respectively. In this ease all second order determinants are zero and hence the matrix is of rank one. If in addition it happens that *a,* b, c, d, e are zero, then the matrix is of rank zero.

4. Special Types of Matrices

a) Row Vector

A set of la quantities arranged in a row is a matrix of order (1 x n). Such a matrix is usually called a row matrix or row vector and is denoted by

$$[A] = \begin{bmatrix} a_1 & a_2 & a_3 & \ldots & a_n \end{bmatrix}$$

b) Column Vector

A set of m quantities arranged in a column is a matrix of order (m x 1). Such a matrix is called a column matrix or column vector and is denoted by

$$\{A\} = \begin{bmatrix} a_1 \\ a_2 \\ a_3 \\ \cdot \\ \cdot \\ \cdot \\ \cdot \\ a_m \end{bmatrix}$$

c) Null – Matrix

Any matrix of order (m x n) with all its elements equal to zero is called a null matrix of order (m x n)

d) Square Matrix

A square matrix has the same number of rows as columns, for example

$$A = \begin{bmatrix} a_{11} & a_{12} & \cdots & a_{1n} \\ a_{21} & a_{22} & \cdots & a_{2n} \\ \cdot & \cdot & & \cdot \\ a_{n1} & a_{n2} & \cdots & a_{nn} \end{bmatrix}$$

is a square matrix of order (n x n). The diagonal containing the elements all
a_{11}, a_{22}, a_{nn}
is called the leading diagonal and the sum of these elements is called the trace (or spur) of the matrix. This sum is usually denoted by Tr (or Sp) for short, for example if

$$A = \begin{bmatrix} 1 & -2 & 3 \\ 2 & -3 & 4 \\ 5 & 6 & 8 \end{bmatrix}$$

then

Tr A (or sp A) = 1 - 3 + 8 = 6

e) Diagonal Matrix

A square matrix with zero elements everywhere except in the leading diagonal is called a diagonal matrix. In other words, if a_{ik} are the elements of a diagonal matrix we must have a_{ik}= 0 for i ≠ k.

A typical (4 x 4) diagonal matrix is

$$A = \begin{bmatrix} 1 & 0 & 0 & 0 \\ 0 & 3 & 0 & 0 \\ 0 & 0 & 5 & 0 \\ 0 & 0 & 0 & 2 \end{bmatrix}$$

Clearly all diagonal matrices of the same order commute under multiplication.

f) Unit Matrix

The unit matrix is a diagonal matrix with all its diagonal elements equal to unity. It is usual to denote such matrices by the letter I; for example the (3x 3) unit matrix is

$$I = \begin{bmatrix} 1 & 0 & 0 \\ 0 & 1 & 0 \\ 0 & 0 & 1 \end{bmatrix}$$

In general, if A is a square matrix of order (m x m) and I is the unit matrix of the same order then

$I A = A I = A$

Likewise

$$I = I^2 = I^3 = \ldots\ldots = I^k \ldots\ldots,$$

where k is any positive integer. In fact multiplying any matrix by a unit matrix leaves the matrix unchanged provided the product is defined. For example

$$\begin{bmatrix} 1 & 0 & 0 \\ 0 & 1 & 0 \\ 0 & 0 & 1 \end{bmatrix} \begin{bmatrix} 2 \\ 3 \\ 5 \end{bmatrix} = \begin{bmatrix} 2 \\ 3 \\ 5 \end{bmatrix}$$

Also if $A = \begin{bmatrix} 1 & 2 & 3 \\ 0 & 1 & 2 \\ 3 & 0 & 1 \end{bmatrix}$

then

$$I A = \begin{bmatrix} 1 & 0 & 0 \\ 0 & 1 & 0 \\ 0 & 0 & 1 \end{bmatrix} \begin{bmatrix} 1 & 2 & 3 \\ 0 & 1 & 2 \\ 3 & 0 & 1 \end{bmatrix} = \begin{bmatrix} 1 & 2 & 3 \\ 0 & 1 & 2 \\ 3 & 0 & 1 \end{bmatrix} = A$$

$$A I = \begin{bmatrix} 1 & 2 & 3 \\ 0 & 1 & 2 \\ 3 & 0 & 1 \end{bmatrix} \begin{bmatrix} 1 & 0 & 0 \\ 0 & 1 & 0 \\ 0 & 0 & 1 \end{bmatrix} = \begin{bmatrix} 1 & 2 & 3 \\ 0 & 1 & 2 \\ 3 & 0 & 1 \end{bmatrix} = A$$

g) Determinant of a Matrix

The determinant of a matrix A is only defined when A is square and is then just the determinant of the matrix elements. It is usually denoted by |A| or det A.
For example if

$$A = \begin{bmatrix} 1 & 0 & 1 \\ 2 & 3 & 1 \\ 3 & 1 & 2 \end{bmatrix}$$

then

$$|A| = \begin{bmatrix} 1 & 0 & 1 \\ 2 & 3 & 1 \\ 3 & 1 & 2 \end{bmatrix} = -2$$

When a square matrix is such that its determinant is zero it is called a singular matrix (otherwise it is non - singular or regular). For example, the matrix

$$\begin{bmatrix} 1 & 2 & 1 \\ 3 & -1 & 3 \\ 5 & 7 & 5 \end{bmatrix}$$

is singular since its determinant vanishes in virtue of two columns being identical.

h) Transposed Matrix

In the case of determinants we have seen that interchanging rows and columns leaves the value of that determinant unaltered. However, if the rows and columns of a matrix are interchanged a new matrix called the transposed matrix is obtained. For example if A is a (3 x 2) matrix given by

$$A = \begin{bmatrix} a_{11} & a_{12} \\ a_{21} & a_{22} \\ a_{31} & a_{32} \end{bmatrix}$$

then its transpose (denoted by A') is the (2 x 3) matrix

$$A' = \begin{bmatrix} a_{11} & a_{21} & a_{31} \\ a_{12} & a_{22} & a_{32} \end{bmatrix}$$

clearly the transpose of a column vector, say

$$\{A\} = \begin{bmatrix} a_1 \\ a_2 \\ a_3 \end{bmatrix}$$

is a row vector since

$$\{A\}'\{A\} = [A][A]' = a_1^2 + a_2^2 + a_3^2$$

Similarly the transpose of a the row vector |A| is the column vector {A}. It follows that

$$\{A\}'\{A\} = [A][A]' = a_1^2 + a_2^2 + a_3^2$$

In general we see that if A is **of** order (m x n) then A' is of order (n x m), and hence A and A' are conformable to both products A A' and A' A (i.e., both products exist but are of different orders unless A is square).

We now show that if A and B are two matrices conformable to the product AB = C, then

C'= (AB)' = B' A'

Suppose A is of order (m x p) with elements a_{ik} and B is of order (p x n) with elements b_{ik}. Then C is a matrix of order (in x n) with elements c_{ik} given by:

$$c_{ik} = \sum_{s=1}^{p} a_{is} b_{sk}$$

consequently

$$c'_{ik} = c = \sum_{s=1}^{p} a_{ks} b_{si} = \sum_{s=1}^{p} b'_{is} a'_{sk}$$

and therefore

C' = B' A'

Similarly we may show that

(ABC)' = C' B' A'

and so on for a finite number of matrices . In other words, in taking the transpose of matrix products the order of the matrices forming the product must he reversed. For example if

$$A = \begin{bmatrix} 1 & 2 \\ 3 & 4 \end{bmatrix}, B = \begin{bmatrix} 2 \\ 1 \end{bmatrix}$$

then

$$A' = \begin{bmatrix} 1 & 3 \\ 2 & 4 \end{bmatrix}, B' = \begin{bmatrix} 2 & 1 \end{bmatrix}$$

and $AB = \begin{bmatrix} 4 \\ 10 \end{bmatrix}$, $B'A' = \begin{bmatrix} 4 & 10 \end{bmatrix}$

clearly (AB)' = B' A' as required.

١٨٨ 1972

The transposed matrix : trans نقل نقل وضع
pose وضع

The transposed matrix of a given matrix is a matrix obtained by interchanging rows & columns.

e.g.

if $A = \begin{bmatrix} 2 & 3 & 4 \\ 5 & -4 & 6 \end{bmatrix}$ Then its transposed

is $A' = \begin{bmatrix} 2 & 5 \\ 3 & -4 \\ 4 & 6 \end{bmatrix}$

If $A = A'$ then the matrix must be square & the elements about the principle diagonal must be equal.

i.e $a_{ij} = a_{ji}$ عناصر متناظرة حول الخط
المستقيم القطري الرئيسي.

e.g. The matrix $\begin{bmatrix} a & h & g \\ h & b & F \\ g & F & c \end{bmatrix}$ principle diagonal

If $A = A'$ the matrix is said to be symmetrical.
On the other hand if $A = -A'$ the matrix is said to be skew symmetrical)

e.g.:-

$$A = \begin{bmatrix} 0 & a & b \\ -a & 0 & c \\ -b & -c & 0 \end{bmatrix}$$

حيث ان (-) و (-) ضرب
كل العناصر تكون المحددة.

$$A' = \begin{bmatrix} 0 & -a & -b \\ a & 0 & -c \\ b & c & 0 \end{bmatrix} = -A$$

we notice that $-a_{ij} = 0$ if $i = j$ (i.e. all elements in the principle diagonal are zero & $a_{ij} = -a_{ji}$ $i \neq j$

i) Symmetric and Skew-symmetric Matrices

A symmetric matrix A is square and has elements a_{ik} such that $a_{ik} = a_{ki}$. In other words, the elements above and below the leading diagonal are mirror images of each other. For example, the matrix

$$A = \begin{bmatrix} 1 & a & b \\ a & 3 & c \\ b & c & 2 \end{bmatrix}$$

is symmetric. Clearly every symmetric matrix is its own transpose in that A A'. Conversely any matrix which is its own transpose is necessarily symmetric. A skew-symmetric matrix is also square but has elements *a* such that $a_{ik} = -a_{ki}$ from which it follows that the leading diagonal has zero elements.

A typical skew symmetric matrix is

$$A = \begin{bmatrix} 0 & 4 & -3 \\ -4 & 0 & 1 \\ 3 & -1 & 0 \end{bmatrix}$$

Such matrices are the negatives of their own transposes (i.e., A = - A')

Any square matrix A can be resolved into the sum of a symmetric matrix $\underline{A}$ and a skew - symmetric matrix $\underset{\smile}{A}$ since if the elements of A are a_{ik} then

$$a_{ik} = (a_{ik} + a_{ki})/2 + (a_{ik} - a_{ki})/2$$

Or
$A = \underline{A} + \underset{\smile}{A}$

For example, if

$$A = \begin{bmatrix} 3 & 1 & 2 \\ 2 & 1 & 3 \\ 1 & 2 & 1 \end{bmatrix}$$

then

$$A = \begin{bmatrix} 3 & \frac{3}{2} & \frac{3}{2} \\ \frac{3}{2} & 1 & \frac{5}{2} \\ \frac{3}{2} & \frac{5}{2} & 1 \end{bmatrix}$$

and

$$A_v = \begin{bmatrix} 0 & -\frac{1}{2} & \frac{1}{2} \\ \frac{1}{2} & 0 & \frac{1}{2} \\ -\frac{1}{2} & -\frac{1}{2} & 0 \end{bmatrix}$$

then

$$A = \begin{bmatrix} 3 & \frac{3}{2} & \frac{3}{2} \\ \frac{3}{2} & 1 & \frac{5}{2} \\ \frac{3}{2} & \frac{5}{2} & 1 \end{bmatrix} + \begin{bmatrix} 0 & -\frac{1}{2} & \frac{1}{2} \\ \frac{1}{2} & 0 & \frac{1}{2} \\ -\frac{1}{2} & -\frac{1}{2} & 0 \end{bmatrix}$$

j) Complex and Hermitian Matrices

If A is any matrix of order (m x n) with complex elements a_{ik}, then the complex conjugate A* of A is found by taking the complex conjugates of all the elements. For example, if

$$A = \begin{bmatrix} 3+2i & 4 & -i \\ 5 & 2-i & 1+i \end{bmatrix}$$

then

$$A^* = \begin{bmatrix} 3-2i & 4 & i \\ 5 & 2+i & 1-i \end{bmatrix}$$

Clearly (A*) * = A

A Hermitian matrix is a square matrix which is unchanged by taking the transpose of its complex conjugate, that is, A is Hermitian if

$$(A^*)' = A$$

For example if

$$A = \begin{bmatrix} 1 & 1+i \\ 1-i & 3 \end{bmatrix}$$

then

$$A^* = \begin{bmatrix} 1 & 1-i \\ 1+i & 3 \end{bmatrix}$$

and consequently

$$(A^*)' = \begin{bmatrix} 1 & 1+i \\ 1-i & 3 \end{bmatrix} = A$$

Hence A is Hermitian. It is obvious that **the diagonal** elements of a Hermitian matrix are always real.

If a square matrix A is such that

(A*)' = — A

it is called a skew-Hermitian.

A Hermitian matrix:-

this is a square matrix unaltered by taking conjugate and transposition. i.e. $\bar{A}' = A$.

e.g Let $A = \begin{bmatrix} a & b+ic & e+if \\ b-ic & d & h+ik \\ e-if & h-ik & g \end{bmatrix}$

diagonal

its conjugate $\bar{A} = \begin{bmatrix} a & b-ic & e-if \\ b+ic & d & h-ik \\ e+if & h+ik & g \end{bmatrix}$

transpose $\bar{A}' = \begin{bmatrix} a & b+ic & e+if \\ b-ic & d & h+ik \\ e-if & h-ik & g \end{bmatrix} = A$

Such a matrix is called a (Hermitian) matrix

On the other hand if $\bar{A}' = -A$ the matrix is called a Skew-Hermitian

$$A = \begin{bmatrix} ia & -b+ic & -e+if \\ b+ic & id & -h+ik \\ e+if & h+ik & ig \end{bmatrix}$$

$$\bar{A}' = \begin{bmatrix} -ia & b-ic & e-if \\ -b-ic & -id & h-ik \\ -e-if & -h-ik & -ig \end{bmatrix} = -A$$

The characteristic eqn of a square matrix :-

* Let A be a square matrix & let I_n be the unit matrix of the same order as A for the matrix $A - \lambda I_n$ the eqn $|A - \lambda I_n| = 0$

The characteristic eqn of the matrix (A) & its roots are called :-

The characteristic rootes of the matrix A or its Latent rootes مميزة or "eigenvalues" قيمة ذاتية

e.g. consider the matrix

$$A = \begin{bmatrix} 2 & 0 & 0 \\ 2 & 3 & 2 \\ 3 & 1 & 2 \end{bmatrix} \quad \therefore \quad I_3 = \begin{bmatrix} 1 & 0 & 0 \\ 0 & 1 & 0 \\ 0 & 0 & 1 \end{bmatrix}$$

$$\therefore A - \lambda I_3 = \begin{bmatrix} 2 & 0 & 0 \\ 2 & 3 & 2 \\ 3 & 1 & 2 \end{bmatrix} - \lambda \begin{bmatrix} 1 & 0 & 0 \\ 0 & 1 & 0 \\ 0 & 0 & 1 \end{bmatrix}$$

$$= \begin{bmatrix} 2-\lambda & 0 & 0 \\ 2 & 3-\lambda & 2 \\ 3 & 1 & 2-\lambda \end{bmatrix} = 0$$

$$|A - \lambda I_3| = 0 = \begin{vmatrix} 2-\lambda & 0 & 0 \\ 2 & 3-\lambda & 2 \\ 3 & 1 & 2-\lambda \end{vmatrix} = 0$$

k) The Adjoint and Reciprocal matrices

If A is the square matrix

$$\begin{bmatrix} a_{11} & a_{12} \cdots\cdots a_{1n} \\ a_{21} & a_{22} \cdots\cdots a_{2n} \\ \cdot & \cdot\cdots\cdots\cdot \\ \cdot & \cdot\cdots\cdots\cdot \\ a_{n1} & a_{n2} \cdots\cdots a_{nn} \end{bmatrix}$$

then its adjoint (usually denoted by adj A) is defined as the transpose of the matrix of its cofactors. In other words, if A_{rs} is the cofactor of the element a_{rs} (i.e., the value of the determinant formed by deleting the row and column in which a_{rs} occurs and attaching a plus or minus sign in accordance with the known rule), then the matrix of the cofactors is the square matrix B (of the same order as A) where

$$B = \begin{bmatrix} A_{11} & A_{12} \cdots\cdots A_{1n} \\ A_{21} & A_{22} \cdots\cdots A_{2n} \\ \cdot & \cdot\cdots\cdots\cdot \\ \cdot & \cdot\cdots\cdots\cdot \\ A_{n1} & A_{n2} \cdots\cdots A_{nn} \end{bmatrix}$$

Consequently

$$\text{adj } A = B' = \begin{bmatrix} A_{11} & A_{21} & \cdots & A_{n1} \\ A_{12} & A_{22} & \cdots & A_{n2} \\ \cdot & \cdot & \cdots & \cdot \\ \cdot & \cdot & \cdots & \cdot \\ A_{1n} & A_{2n} & \cdots & A_{nn} \end{bmatrix}$$

For example, if

$$A = \begin{bmatrix} 1 & 2 & 3 \\ 1 & 3 & 5 \\ 1 & 5 & 12 \end{bmatrix}$$

then the cofactor of a_{11} A_{11} = (3x 12-5 x 5) = 11 and the cofactor of a_{12} is A_{12} = - (12 x 1-5 x 1) = -. 7 and so on. Proceeding in this was• we find that the matrix of cofactors is

$$B = \begin{bmatrix} 11 & -7 & 2 \\ -9 & 9 & -3 \\ 1 & -2 & 1 \end{bmatrix}$$

Hence by definition

$$\text{adj } A = B' = \begin{bmatrix} 11 & -9 & 1 \\ -7 & 9 & -2 \\ 2 & -3 & 1 \end{bmatrix}$$

We now consider the product

$$A\,(\text{adj } A) = \begin{bmatrix} a_{11} & a_{12} & \cdot\cdot & a_{1n} \\ a_{21} & a_{22} & \cdot\cdot & a_{2n} \\ \cdot & \cdot & \cdot\cdot & \cdot \\ & \cdot & \cdot\cdot & \cdot \\ a_{n1} & a_{n2} & \cdot\cdot & a_{nn} \end{bmatrix} \begin{bmatrix} A_{11} & A_{21} & \cdot\cdot & A_{n1} \\ A_{12} & A_{22} & \cdot\cdot & A_{n2} \\ \cdot & \cdot & \cdot\cdot & \cdot \\ \cdot & \cdot & \cdot\cdot & \cdot \\ A_{1n} & A_{2n} & \cdot\cdot & A_{nn} \end{bmatrix}$$

$$= \begin{bmatrix} |A| & 0 & 0 \cdots & 0 \\ 0 & |A| & 0 \cdots & 0 \\ 0 & 0 & |A| \cdots & 0 \\ \cdot & \cdot & \cdot\cdot\cdot\cdot & \cdot \\ 0 & 0 & 0\;0 \cdot\cdot & |A| \end{bmatrix} = |A|\, I$$

where $|A|$ is the determinant of A and I is the unit matrix of the same order as A

It follows that the martix A^{-1} defined by

It follows that the martix A^{-1} defined by

$$A^{-1} = \frac{\text{adj } A}{|A|}$$

has the property that $AA^{-1} = I$

and in this sense it is referred to as the reciprocal (or inverse) matrix of A. It is easily verified that multiplication of matrices and their inverses is commutative in that

$A\,A^{-1} = A^{-1}\,A = I$

Clearly only nonsingular square matrices have reciprocals, since A^{-1} is undefined when $|A| = 0$

Example

(6)

Find A^{-1} if $A = \begin{bmatrix} 1 & 2 & 3 \\ 1 & 3 & 5 \\ 1 & 5 & 12 \end{bmatrix}$

From the last example, we have

$$\text{adj } A = \begin{bmatrix} 11 & -9 & 1 \\ -7 & 9 & -2 \\ 2 & -3 & 1 \end{bmatrix}$$

|A| = 11 - 2x7 + 3x2 = 3

$$A^{-1} = \frac{\text{adj } A}{|A|} = \begin{bmatrix} \frac{11}{3} & -3 & \frac{1}{3} \\ -\frac{7}{3} & 3 & -\frac{2}{3} \\ \frac{2}{3} & -1 & \frac{1}{3} \end{bmatrix}$$

We shall now show how a non - singular square matrix can be transferred from one side of a matrix equation to the other side. Suppose A is a non - singular square matrix and suppose that B and C are two matrices such that AB =C, we shall now show that
$B = A^{-1} C$

For $A^{-1} C = A^{-1} A B = I B = B$

Thus A occupies the first position on L. H. S. and A^{-1} occupies also the first position on R. H. S.

Again suppose that BA = D. We shall show that
$B = D A^{-1}$

For $D A^{-1} = BA A^{-1} = B I = B$

Thus A occupies the second position on L. H. S, and A^{-1} occupies also the second position on R. H. S. Hence we have the following rule:

If a non - singular matrix is transferred from one side of a matrix equation to the other, then its inverse must occupy in the new side a position similar to that which the matrix occupied in the original side.

The Adjugate or adjoint matrix:-

Def:- Let A be a square matrix then its adjugate or adjoint matrix is the matrix obtained by replacing each element by the cofactor but in the transposed position. i.e for example-

If $A = \begin{bmatrix} a_{11} & a_{12} & a_{13} \\ a_{21} & a_{22} & a_{23} \\ a_{31} & a_{32} & a_{33} \end{bmatrix}$

The adj A is

$$\begin{bmatrix} A_{11} & A_{21} & A_{31} \\ A_{12} & A_{22} & A_{32} \\ A_{13} & A_{23} & A_{33} \end{bmatrix}$$

Where the capital letters are the cofactor of the corresponding elements

Note:-

The minor of element in the determinant is the determinant which remains after erasing i.e (eliminating) the row & column through the element.

On the otherhand the cofactor of an element is its minor with a sign prefixed to it according to the following scheme.

For example, in the determinant

$$\begin{vmatrix} 3 & 4 & 5 \\ 6 & 1 & 2 \\ 5 & 6 & 3 \end{vmatrix}$$

$$\begin{matrix} + & - & + & - & + & \cdots \\ + & - & + & - & \cdots \\ + & - & + & - & + & \cdots \\ - & + & - & + & - & \cdots \\ - & - & - & - & \cdots \end{matrix}$$

$$\text{cofactor}(3) = +\begin{vmatrix} 1 & 2 \\ 6 & 3 \end{vmatrix} \qquad \text{Minor}(3) = \begin{vmatrix} 1 & 2 \\ 6 & 3 \end{vmatrix}$$

while

$$\text{cofactor}(6) = -\begin{vmatrix} 4 & 5 \\ 6 & 3 \end{vmatrix} \qquad \text{Minor}(6) = \begin{vmatrix} 4 & 5 \\ 6 & 3 \end{vmatrix}$$

$$\text{cofactor}(4) = -\begin{vmatrix} 6 & 2 \\ 5 & 3 \end{vmatrix} \qquad \text{Minor}(4) = \begin{vmatrix} 6 & 2 \\ 5 & 3 \end{vmatrix}$$

Now let us consider $A(\text{adj} A)$

$$A(\text{adj} A) = \begin{bmatrix} a_{11} & a_{12} & a_{13} \\ a_{21} & a_{22} & a_{23} \\ a_{31} & a_{32} & a_{33} \end{bmatrix} \begin{bmatrix} A_{11} & A_{21} & A_{31} \\ A_{12} & A_{22} & A_{32} \\ A_{13} & A_{23} & A_{33} \end{bmatrix}$$

$$A(\operatorname{adj} A) = \begin{bmatrix} a_{11} & a_{12} & a_{13} \\ a_{21} & a_{22} & a_{23} \\ a_{31} & a_{32} & a_{33} \end{bmatrix} \begin{bmatrix} A_{11} & A_{21} & A_{31} \\ A_{12} & A_{22} & A_{32} \\ A_{13} & A_{23} & A_{33} \end{bmatrix}$$

determinant, zero, zero

$$= \begin{bmatrix} a_{11}A_{11} + a_{12}A_{12} + a_{13}A_{13} & a_{11}A_{21} + a_{12}A_{22} + a_{13}A_{23} & a_{11}A_{31} + a_{12}A_{32} + a_{13}A_{33} \\ a_{21}A_{11} + a_{22}A_{12} + a_{23}A_{13} & a_{21}A_{21} + a_{22}A_{22} + a_{23}A_{23} & a_{21}A_{31} + a_{22}A_{32} + a_{23}A_{33} \end{bmatrix}$$

zero, zero

$$= \begin{bmatrix} a_{11}A_{11} + a_{12}A_{12} + a_{13}A_{13} & a_{11}A_{21} + a_{12}A_{22} + a_{13}A_{23} & a_{11}A_{31} + a_{12}A_{32} + a_{13}A_{33} \\ a_{21}A_{11} + a_{22}A_{12} + a_{23}A_{13} & a_{21}A_{21} + a_{22}A_{22} + a_{23}A_{23} & a_{21}A_{31} + a_{22}A_{32} + a_{23}A_{33} \\ a_{31}A_{11} + a_{32}A_{12} + a_{33}A_{13} & a_{31}A_{21} + a_{32}A_{22} + a_{33}A_{23} & a_{31}A_{31} + a_{32}A_{32} + a_{33}A_{33} \end{bmatrix}$$

$$= \begin{bmatrix} \text{determinant} & \text{zero} & \text{zero} \\ \text{zero} & \text{determinant} & \text{zero} \\ \text{zero} & \text{zero} & \text{determinant} \end{bmatrix}$$

$$= \begin{bmatrix} |A| & 0 & 0 \\ 0 & |A| & 0 \\ 0 & 0 & |A| \end{bmatrix} = |A| \begin{bmatrix} 1 & 0 & 0 \\ 0 & 1 & 0 \\ 0 & 0 & 1 \end{bmatrix}$$

$$\therefore \boxed{A \operatorname{adj} A = |A| I_n}$$

Hence in general we have:-

$$A(\operatorname{adj} A) = (\operatorname{adj} A)A = |A| I_n$$

Where I_n is a unit matrix of order $n \times n$.

Note: • If the elements of a row or a column be multiplied by their cofactors & the results are added we obtain the det^{t}. If the elements of a row det^{t} be multiplied by the cofactor of a second row & the results added we obtain zero

5. Pre Division and Post Division

Let A be a non singular square matrix of order n x n and B is a matrix. We are now in a position to give meaning to B ÷ A by finding the matrix P or Q such that A P = B or Q A = B.

If A P = B then P = A^{-1} B. This implies that B has n rows. This is called a pre-dvision.

If QA = B then Q = B A^{-1}. This implies that B has n columns. This is called a post division

If B is a square matrix of order (n x n) then both P and Q exist but they are in general different.

The Reciprocal or inverse of a non singlar or square matrix:

Def:- The inverse matrix A^{-1} of a non singular square matrix A is its adjugate but with each element devided by |A| thus if:—

If $A = \begin{bmatrix} a_{11} & a_{12} & \cdots & a_{1n} \\ a_{21} & a_{22} & \cdots & a_{2n} \\ \cdots & \cdots & \cdots & \cdots \\ a_{n1} & a_{n2} & \cdots & a_{nn} \end{bmatrix}$ where $|A| \neq 0$ $\therefore A^{-1} = \begin{bmatrix} \frac{A_{11}}{|A|} & \frac{A_{21}}{|A|} & \frac{A_{n1}}{|A|} \\ \frac{A_{12}}{|A|} & \frac{A_{22}}{|A|} & \frac{A_{n2}}{|A|} \\ \frac{A_{1n}}{|A|} & \frac{A_{2n}}{|A|} & \frac{A_{nn}}{|A|} \end{bmatrix}$

now we have already seen that:—

$$A\,(\operatorname{adj} A) = |A|\, I_n$$

$$A\,\frac{(\operatorname{adj} A)}{|A|} = \frac{(\operatorname{adj} A)}{|A|}\,A = I_n$$

$$AA^{-1} = A^{-1}A = I_n$$ قیاساً على الجبر العادی

Now suppose that A is a non singular square matrix & suppose that we have the matrix eqn $AB = C$. where A, B, C are matrices. (Algebra of matrices) to show that.

$\because AB = C$

For $B = A^{-1}C$,

For $A^{-1}C = A^{-1}AB = I_n B = B$

Similarily if $BA = D$ then

$B = DA^{-1}$

For $DA^{-1} = BAA^{-1} = BI = B$

<u>Hence we have the following rule.</u>

If a non singular matrix be transfered from one side of a matrix eqn to the other side then its transferred in the inverse form & occupies on the new side a position similar to that it had on the original side.

example- Solve the n eqns given by. $AX = B$

where

$$A = \begin{bmatrix} a_{11} & a_{12} & \cdots & a_{1n} \\ a_{21} & a_{22} & \cdots & a_{2n} \\ a_{n1} & a_{n2} & \cdots & a_{nn} \end{bmatrix}, \quad X = \begin{bmatrix} x_1 \\ x_2 \\ x_n \end{bmatrix}$$

$$B = \begin{bmatrix} b_1 \\ b_2 \\ b_n \end{bmatrix}$$ writting in a full we get

Sol^n

$$\begin{bmatrix} a_{11} & a_{12} & \cdots & a_{1n} \\ a_{21} & a_{22} & \cdots & a_{2n} \\ a_{n1} & a_{n2} & \cdots & a_{nn} \end{bmatrix} \begin{bmatrix} x_1 \\ x_2 \\ x_n \end{bmatrix} = \begin{bmatrix} b_1 \\ b_2 \\ b_n \end{bmatrix} \longrightarrow (1)$$

$$\therefore \left.\begin{aligned} a_{11}x_1 + a_{12}x_2 + \cdots + a_{1n}x_n &= b_1 \\ a_{21}x_1 + a_{22}x_2 + \cdots + a_{2n}x_n &= b_2 \\ \text{---------} \\ a_{n1}x_1 + a_{n2}x_2 + \cdots + a_{nn}x_n &= b_n \end{aligned}\right] \longrightarrow (2)$$

Writting the matrix eq^n in the inverse form we get

$$X = A^{-1}B$$

$$\begin{bmatrix} x_1 \\ x_2 \\ x_n \end{bmatrix} = \begin{bmatrix} \frac{A_{11}}{|A|} & \frac{A_{21}}{|A|} & \frac{A_{n1}}{|A|} \\ \frac{A_{12}}{|A|} & \frac{A_{22}}{|A|} & \frac{A_{n2}}{|A|} \\ \frac{A_{1n}}{|A|} & \frac{A_{2n}}{|A|} & \frac{A_{nn}}{|A|} \end{bmatrix} \begin{bmatrix} b_1 \\ b_2 \\ b_n \end{bmatrix}$$

$$x_1 = \frac{b_1 A_{11} + b_2 A_{21} + \cdots + b_n A_{n1}}{|A|} = \frac{\begin{vmatrix} b_1 & a_{12} & a_{1n} \\ b_2 & a_{22} & a_{2n} \\ b_n & a_{n2} & a_{nn} \end{vmatrix}}{\begin{vmatrix} a_{11} & a_{12} & \cdots & a_{1n} \\ a_{21} & a_{22} & \cdots & a_{2n} \\ a_{n1} & a_{n2} & & a_{nn} \end{vmatrix}}$$

$$x_2 = \frac{b_1 A_{12} + b_2 A_{22} + \cdots + b_n A_{n2}}{|A|} = \frac{\begin{vmatrix} a_{11} & b_1 & a_{1n} \\ a_{21} & b_2 & a_{2n} \\ a_{n1} & b_n & a_{nn} \end{vmatrix}}{\begin{vmatrix} a_{11} & a_{12} & a_{1n} \\ a_{21} & a_{22} & a_{2n} \\ a_{n1} & a_{n2} & a_{nn} \end{vmatrix}}$$

$$x_3 = \frac{b_1 A_{1n} + b_2 A_{2n} + \cdots + b_n A_{nn}}{|A|} = \frac{\begin{vmatrix} a_{11} & a_{12} & b_1 \\ a_{21} & a_{22} & b_2 \\ a_{n1} & a_{n2} & b_n \end{vmatrix}}{\begin{vmatrix} a_{11} & a_{12} & a_{1n} \\ a_{21} & a_{22} & a_{2n} \\ a_{n1} & a_{n2} & a_{nn} \end{vmatrix}}$$

Similtarily for the other (unknowns) this rule is known as (Cramer's rule)

<u>Orthogonal matrices</u> :-

$$X_3 = \frac{b_1 A_{1n} + b_2 A_{2n} + \cdots + b_n A_{nn}}{|A|} = \frac{\begin{vmatrix} a_{11} & a_{12} & b_1 \\ a_{21} & a_{22} & b_2 \\ a_{n1} & a_{n2} & b_n \end{vmatrix}}{\begin{vmatrix} a_{11} & a_{12} & a_{1n} \\ a_{21} & a_{22} & a_{2n} \\ a_{n1} & a_{n2} & a_{nn} \end{vmatrix}}$$

Similarily for the other (unknowns) this rule is known as (Cramer's rule)

6. Orthogonal Matrices

A square matrix A is said to he orthogonal if A' A = I

$\therefore$

AA' = A I A^{-1} = A A^{-1} = I

Hence for an orthogonal matrix we have A' A. = A A' = I

Example

(7)

show that $A = \begin{bmatrix} \cos\theta & -\sin\theta \\ \sin\theta & \cos\theta \end{bmatrix}$ is an orthogonal matrix

$$A'A = \begin{bmatrix} \cos\theta & \sin\theta \\ -\sin\theta & \cos\theta \end{bmatrix} \begin{bmatrix} \cos\theta & -\sin\theta \\ \sin\theta & \cos\theta \end{bmatrix}$$

$$= \begin{bmatrix} \cos^2\theta + \sin^2\theta & -\cos\theta\sin\theta + \sin\theta\cos\theta \\ -\sin\theta\cos\theta + \cos\theta\sin\theta & \sin^2\theta + \cos^2\theta \end{bmatrix}$$

$$= \begin{bmatrix} 1 & 0 \\ 0 & 1 \end{bmatrix} = I$$

Hence A is orthogonal.

7. The characteristic Equation of a Matrix

If A is a square matrix of order (n x n) and I is the unit matrix of the same order, then the matrix

$$B = A - \lambda I$$

is called the characteristic matrix of A, λ being a parameter. For example if

$$A = \begin{bmatrix} 1 & 0 & 1 \\ 0 & 2 & 0 \\ 1 & 1 & 3 \end{bmatrix}$$

then

$$B = \begin{bmatrix} 1 & 0 & 1 \\ 0 & 2 & 0 \\ 1 & 1 & 3 \end{bmatrix} - \lambda \begin{bmatrix} 1 & 0 & 0 \\ 0 & 1 & 0 \\ 0 & 0 & 1 \end{bmatrix}$$

$$= \begin{bmatrix} 1-\lambda & 0 & 1 \\ 0 & 2-\lambda & 0 \\ 1 & 1 & 3-\lambda \end{bmatrix}$$

The equation

$$|B| = | A - \lambda I | = 0$$

is called the characteristic equation of A and is in general an equation of the n th degree in λ. The n roots of this equation are called the characteristic or latent roots (or eigen values) of A.

For example, the characteristic equation

of the above matrix A is obtained by equating the determinant $|A - \lambda I|$ to zero. This gives

$$(\lambda - 2)(\lambda^2 - 4\lambda + 2) = 0$$

whose solutions (the eigenvalues) are

$$\lambda = 2,\ 2 \pm \sqrt{2}$$

8. Solution of Linear Equations

Consider the set of n linear equations in the n unknowns $x_1, x_2, \ldots, x_n$

$$\begin{aligned} a_{11}x_1 + a_{12}x_2 + \cdots + a_{1n}x_n &= h_1, \\ a_{21}x_1 + a_{22}x_2 + \cdots + a_{2n}x_n &= h_2, \\ \cdot \quad \cdot \quad \cdot \quad & \cdot \quad \cdot \\ a_{n1}x_1 + a_{n2}x_2 + \cdots + a_{nn}x_n &= h_n \end{aligned}$$

where the coefficients a_{ik} and the h_i are known constants. Using the concept of matrix multiplication, it is clear that these equations can be written in matrix form as

$$\begin{bmatrix} a_{11} & a_{12} & \cdots & a_{1n} \\ a_{21} & a_{22} & \cdots & a_{2n} \\ \cdot & \cdot & \cdots & \cdot \\ \cdot & \cdot & \cdots & \cdot \\ \cdot & \cdot & \cdots & \cdot \\ a_{n1} & a_{n2} & \cdots & a_{nn} \end{bmatrix} \begin{bmatrix} x_1 \\ x_2 \\ \cdot \\ \cdot \\ \cdot \\ x_n \end{bmatrix} = \begin{bmatrix} h_1 \\ h_2 \\ \cdot \\ \cdot \\ \cdot \\ h_n \end{bmatrix}$$

Or

$A\{x\} = \{h\}$

where

$$A = \begin{bmatrix} a_{11} & a_{12} & \dots & a_{1n} \\ a_{21} & a_{22} & \dots & a_{2n} \\ . & . & & . \\ . & . & & . \\ a_{n1} & a_{n2} & & a_{nn} \end{bmatrix}, \{x\} = \begin{bmatrix} x_1 \\ x_2 \\ . \\ . \\ x_n \end{bmatrix}$$

and

$$\{h\} = \begin{bmatrix} h_1 \\ h_2 \\ . \\ . \\ h_n \end{bmatrix}$$

If we now form the reciprocal matrix A^{-1} of A and apply it to both sides of the equation $A\{x\} = \{h\}$ we obtain

$$A^{-1} A\{x\} = A^{-1}\{h\}$$

which gives (since $A^{-1}A = I$)

$$I\{x\} = \{x\} = A^{-1}\{h\} = \frac{\text{adj } A}{|A|}\{h\}$$

which is the solution of the matrix equation $A\{x\} = \{h\}$ and enables the values of $x_1, x_2, \dots, x_n$ to be found directly by evaluating $A^{-1}\{h\}$

Example ((8)

Solve the equations

$$x_1 + x_2 + x_3 = 6$$

$$x_1 + 2x_2 + 3x_3 = 14$$

$$x_1 + 4x_2 + 9x_3 = 36$$

$$A = \begin{bmatrix} 1 & 1 & 1 \\ 1 & 2 & 3 \\ 1 & 4 & 9 \end{bmatrix}, \{x\} = \begin{bmatrix} x_1 \\ x_2 \\ x_3 \end{bmatrix} \text{ and } \{h\} = \begin{bmatrix} 6 \\ 14 \\ 36 \end{bmatrix}$$

$$|A| = (18 - 12) - (9 - 3) + (4 - 2) = 2$$

$$\text{Adj } A = \begin{bmatrix} 6 & -5 & 1 \\ -6 & 8 & -2 \\ 2 & -3 & 1 \end{bmatrix}$$

$$A^{-1} = \frac{\text{adj } A}{|A|} = \begin{bmatrix} 3 & -\frac{5}{2} & \frac{1}{2} \\ -3 & 4 & -1 \\ 1 & -\frac{3}{2} & \frac{1}{2} \end{bmatrix}$$

$$\{x\} = A^{-1}\{h\}$$

$$\therefore \begin{bmatrix} x_1 \\ x_2 \\ x_3 \end{bmatrix} = \begin{bmatrix} 3 & -\frac{5}{2} & \frac{1}{2} \\ -3 & 4 & -1 \\ 1 & -\frac{3}{2} & \frac{1}{2} \end{bmatrix} \begin{bmatrix} 6 \\ 14 \\ 36 \end{bmatrix} = \begin{bmatrix} 1 \\ 2 \\ 3 \end{bmatrix}$$

$$\therefore x_1 = 1,\ x_2 = 2,\ x_3 = 3$$

9. Consistent and Inconsistent systems of Equations

A set of equations that have at least one common solution is said to be a consistent set of equations. A set for which there exists no common solution is called an inconsistent set.

The question of consistency is frequently of practical importance. For example in setting up problems in electrical net-works, there are often more conditions than there are variables. This leads to a system in which there are more equations than there are unknowns. It is important to have a method for testing whether all the conditions can be satisfied simultaneously.

Consider a system of m linear equations in n unknowns,

$$
\begin{aligned}
a_{11} x_1 + a_{12} x_2 + \ldots\ldots + a_{1n} x_n &= h_1 \\
a_{21} x_1 + a_{22} x_2 + \ldots\ldots + a_{2n} x_n &= h_2 \\
\ldots\ldots\ldots\ldots\ldots\ldots\ldots\ldots\ldots \\
a_{m1} x_1 + a_{m2} x_2 + \ldots\ldots + a_{mn} x_n &= h_m
\end{aligned}
$$

where at least one $h_i \neq 0$

From the matrix of coefficients, *namely*

$$
\begin{bmatrix}
a_{11} & a_{12} & \cdots & a_{1n} \\
a_{21} & a_{22} & \cdots & a_{2n} \\
\cdot & \cdot & & \cdot \\
a_{m1} & a_{m2} & \cdots & a_{mn}
\end{bmatrix}
$$

Let its rank be r. Form the augmented matrix by adjoining to the matrix of the coefficients the column in the right band side of the given equations, i.e.,

$$
\begin{bmatrix}
a_{11} & a_{12} & a_{1n} & \cdots & h_1 \\
a_{21} & a_{22} & a_{2n} & \cdots & h_2 \\
\cdot & \cdot & \cdot & & \cdot \\
a_{m1} & a_{m2} & a_{mn} & \cdots & h_m
\end{bmatrix}
$$

Let its rank be r'.

Then, we have the following fundamental rule

I) Unique solution: if $r = r' = n$
II) More than one solution: if $r = r' < n$

In this case we can give $n - r$ of the unknowns arbitrary values and then express the remaining r unknowns in terms of these. The r unknowns, which are expressed in terms of the others must be associated with some non vanishing determinant of order r

III) No solution if $r' > r$

Since r can not exceed r' since all determinants in the matrix of coefficients are included in the augmented matrix.

We can summarize as follows:

i) The equations are consistent if $r = r'$
ii) The equations are inconsistent $r \neq r'$

We illustrate by the case $n = 3$

Suppose we are given three linear equations in three unknowns x, y, z namely:

$$a_1 x + b_1 y + c_1 z = d_1$$

$$a_2 x + b_2 y + c_2 z = d_2$$

$$a_3 x + b_3 y + c_3 z = d_3$$

The three equations represent geometrically three planes.

Case I: Unique solution

The matrix of coefficients is

$$\begin{bmatrix} a_1 & b_1 & c_1 \\ a_2 & b_2 & c_2 \\ a_3 & b_3 & c_3 \end{bmatrix} \text{ of rank } r$$

The augmented matrix is

$$\begin{bmatrix} a_1 & b_1 & c_1 & d_1 \\ a_2 & b_2 & c_2 & d_2 \\ a_3 & b_3 & c_3 & d_3 \end{bmatrix} \text{ of rank } r'$$

Since there in a unique solutions the three planes intersect in one point which can be obtained by solving the three equations. The necessary condition for having a single solution is that:

$$\begin{bmatrix} a_1 & b_1 & c_1 \\ a_2 & b_2 & c_2 \\ a_3 & b_3 & c_3 \end{bmatrix} \neq 0$$

This weans that r = 3. Also r' = 3

$\therefore$

$r = r' = n$

The solution is quickly obtained by using matrices as in the previous example.

<u>Case II:</u> More than one solution.

This occurs in the following cases:
(i) The three planes have a common line of intersection
In this case

$$a_3 x + b_3 y + c_3 z - d_3 = (a_1 x + b_1 y + c_1 z - d_1) + \lambda(a_2 x + b_2 y + c_2 z - d_2)$$

$$\text{i-e } \frac{a_3}{a_1+\lambda a_2} = \frac{b_3}{b_1+\lambda b_2} = \frac{c_3}{c_1+\lambda c_2} = \frac{d_3}{d_1+\lambda d_2}$$

a3 x b3 y z — d3 = *(al* x hl y cl z — dl)
+2, *(a, x + b,,* y z — d2)

The matrix of coefficients now becomes

$$\begin{bmatrix} a_1 & b_1 & c_1 \\ a_2 & b_2 & c_2 \\ a_1+\lambda a_2 & b_1+\lambda b_2 & c_1+\lambda c_2 \end{bmatrix}$$

$$\begin{vmatrix} a_1 & b_1 & c_1 \\ a_2 & b_2 & c_2 \\ a_1+\lambda a_2 & b_1+\lambda b_2 & c_1+\lambda c_2 \end{vmatrix} = \begin{vmatrix} a_1 & b_1 & c_1 \\ a_2 & b_2 & c_2 \\ a_1 & b_1 & c_1 \end{vmatrix}$$

$$+ \lambda \begin{vmatrix} a_1 & b_1 & c_1 \\ a_2 & b_2 & c_2 \\ a_2 & b_2 & c_2 \end{vmatrix} = 0$$

$\therefore$

r -= 2 and similarly r' = 2

$\therefore$

r = r' < n

In this ease we can give z any value and solve for x and y in any two of the given equations.

ii) Two of the three planes coincide say the first two planes

$$\therefore \frac{a_1}{a_2} = \frac{b_1}{b_2} = \frac{c_1}{c_2} = \frac{d_1}{d_2}$$

Then r = 2, r' = 2

$\therefore$

r = r' < n

In this case also, we can give z any value and solve for x and y in the two different equations.

iii) The three planes coincide, then r = 1, r' = 1

$\therefore$

$r = r' < n$

In this case we can give x and y any arbitrary values and then solve for z.

Case III: No solution

This occurs in the following cases:
i) Two of the planes are parallel.

If the first two planes are parallel then $\frac{a_1}{a_2} = \frac{b_1}{b_2} = \frac{c_1}{c_2}$

therefore r = 2, r' = 3. $\therefore r' > r$

ii) The three planes are parallel

$$\therefore \frac{a_1}{a_2} = \frac{b_1}{b_2} = \frac{c_1}{c_2}, \quad \frac{a_2}{a_3} = \frac{b_2}{b_3} = \frac{c_2}{c_3}$$

$\therefore$ r = 1, r' = 2, r' > r

iii) The line of intersection of two planes is parallel to the third plane.

If the line of intersection of the first two planes is parallel to the third plane, we have

$$\frac{l}{b_1c_2 - b_2c_1} = \frac{m}{c_1a_2 - c_2a_1} = \frac{n}{a_1b_2 - a_2b_1}$$

where *l, m, n* are the direction ratios of the line of intersection of the first two planes.

Since this line is parallel to the third plane, the normal to the third plane is perpendicular to this line, hence,

$$a_3(b_1c_2 - b_2c_1) + b_3(c_1a_2 - c_2a_1) + c_3(a_1b_2 - a_2b_1) = 0$$

$$\therefore \begin{vmatrix} a_1 & b_1 & c_1 \\ a_2 & b_2 & c_2 \\ a_3 & b_3 & c_3 \end{vmatrix} = 0$$

Hence $r = 2$, $r' = 3$ i-e. $r' > r$

Example (9)
Show that the following equations are consistent if and only if k = 9

$$2x + 3y = 1,$$
$$x - 2y = 4,$$
$$4x - y = k$$

The matrix of coefficients is $\begin{bmatrix} 2 & 3 \\ 1 & -2 \\ 4 & -1 \end{bmatrix}$,

The augmented matrix is $\begin{bmatrix} 2 & 3 & 1 \\ 1 & -2 & 4 \\ 4 & -1 & k \end{bmatrix}$

r = 2, hence r' should be equal 2 in order that the equations may be consistent

$$\therefore \begin{vmatrix} 2 & 3 & 1 \\ 1 & -2 & 4 \\ 4 & -1 & k \end{vmatrix} = 0$$

$\therefore$

$$2(-2k + 4) - 3(k - 16) - 1 + 8 = 0$$

$\therefore$

$$k = 9$$

Example (10)
Solve the system

$x - y + 2z = 3$
$x + y - 2z = 1$
$x + 3y - 6z = -1$

The matrix of coefficients is $\begin{bmatrix} 1 & -1 & 2 \\ 1 & 1 & -2 \\ 1 & 3 & -6 \end{bmatrix}$

$r = 2$ since the determinant of the above matrix $= 0$

Augmented matrix is $\begin{bmatrix} 1 & -1 & 2 & 3 \\ 1 & 1 & -2 & 1 \\ 1 & 3 & -6 & -1 \end{bmatrix}$

r' = 2

$\therefore$

$r = r' < n$

Hence the equations arc consistent, we can give z any value and solve for x and in any two equations. If $z = k$, the equations to be solved are

$x - y = 3 - 2k,$
$x + y = 1 + 2k,$
$x + 3y = -1 + 6k$

Solving the first two for x and y gives $x = \mathbf{2}$ and $y = 2k - 1$. These values are seen to satisfy the third equation.
Therefore the solutions x — 2, $y = 2k - 1$, $z = k$ satisfy the original system for all values of k.

Example (11)
Examine the equations

$2x + y - z = 1,$
$x - 2y + z = 3,$
$4x - 3y + z = 5$

The matrix of coefficients is $\begin{bmatrix} 2 & 1 & -1 \\ 1 & -2 & 1 \\ 4 & -3 & 1 \end{bmatrix}$

$r = 2$ since the determinant of the above matrix $= 0$

The augmented matrix is $\begin{bmatrix} 2 & 1 & -1 & 1 \\ 1 & -2 & 1 & 3 \\ 4 & -3 & 1 & 5 \end{bmatrix}$

$r' = 3$, since $\begin{bmatrix} 1 & -1 & 1 \\ -2 & 1 & 3 \\ -3 & 1 & 5 \end{bmatrix} \neq 0$

$r' > r$ so that the system of equations is inconsistent.

Important transformation:—

m linear simultaneous eq^{n} in n unknowns:-

consider the eq^{ns}. عدد المجاهيل لا يساوى المعادلات.

$$a_{11} X_1 + a_{12} X_2 + a_{13} X_3 + \cdots + a_{1n} X_n = b_1$$
$$a_{21} X_1 + a_{22} X_2 + a_{23} X_3 + \cdots + a_{2n} X_n = b_2$$
$$- - - - - - - - - - - - - -$$
$$a_{m1} X_1 + a_{m2} X_2 + a_{m3} X_3 + \cdots + a_{mn} X_n = b_m$$

If there exists a <u>set</u> مجموعة of values of $x_1, x_2, \ldots, x_n$ which satisfy all eqns then we say that the eqns are <u>consistent</u> متوافق, this includes two cases:-

i) Unique solution.

ii) More than one solution.] consistent.

On the other hand if there exists no set of values which satisfy all eqns then we say that the eqns are Inconsistent غير متوافقة.

To distinguish betn the three cases we adopt the following fundamental rule:-

Fundamental rule to distinguish betn

1- consistent eqns — unique soln / More than one soln

2- inconsistent eqns

eqns

$$a_{11}x_1 + a_{12}x_2 + \cdots + a_{12}x_n = b_1$$

$$a_{21}x_1 + a_{22}x_2 + \cdots + a_{2n}x_n = b_2$$

$$\cdots$$

$$a_{m1}x_1 + a_{m2}x_2 + \cdots + a_{mn}x_n = b_m$$

(i)" Form the matrix of the coefficients :-

$$\begin{bmatrix} a_{11} & a_{12} \cdots & a_{1n} \\ a_{21} & a_{22} \cdots & a_{2n} \\ \cdots & - - & - - \\ a_{m1} & a_{m2} \cdots & a_{mn} \end{bmatrix}$$

and let its rank be (r).

(ii) We form the augmented matrix [illegible]. obtained by adjoining [illegible] the rightest [illegible] column to the right to the matrix of coefficients. i.e the augmented matrix is

$$\begin{bmatrix} a_{11} & a_{12} \cdots & a_{1n} & b_1 \\ a_{21} & a_{22} \cdots & a_{2n} & b_2 \\ a_{m1} & a_{m2} \cdots & a_{mn} & b_m \end{bmatrix}$$

and let its rank be r'.

(I) Unique solⁿ:- $r = r' = n$ where n is the number of unknowns.

(II) More than one solⁿ:- $r = r' < n$

(III) No solⁿ:- $r' > r$

Since r cannot be greater than r' the matrix of the cofft being included the augment matrix. Then we can summarize the above rules as follows:-

Eqns consistent $r = r'$
Eqns inconsistent $r \neq r'$

ex:- 1- Show that the following eqns are consistent if only $k = 9$)

$2x + 3y = 1$
$x - 2y = 4$
$4x - y = k$

matrix of cofft

$$\begin{bmatrix} 2 & 3 \\ 1 & -2 \\ 4 & -1 \end{bmatrix}$$

Augmented matrix

$$\begin{bmatrix} 2 & 3 & 1 \\ 1 & -2 & 4 \\ 4 & -1 & k \end{bmatrix}$$

The highest determinant which can be formed up of the matrix of the coeff^{ts} of order $\underline{\underline{2}}$ we have at least one det$^{\underline{n}}$ of the second order. which is not zero.

$$\begin{vmatrix} 2 & 3 \\ 1 & -2 \end{vmatrix} \neq 0 \quad \therefore \quad \underline{\underline{r = 2}}$$

In order that the eq$^{\underline{ns}}$ may be consistent then $r = r'$ & hence

$$\begin{vmatrix} 2 & 3 & 1 \\ 1 & -2 & 4 \\ 4 & -1 & k \end{vmatrix} = 0 \quad \therefore \quad k = 9$$

i.e the third ($3^{\underline{rd}}$) order det$^{\underline{n}}$ must equal zero.

---

ex.2 Show that the following eq$^{\underline{ns}}$ have a unique sol$^{\underline{n}}$ & find this sol$^{\underline{n}}$.

$$x + 2y + 3z = 2$$
$$4x + 6y + 7z = 5$$
$$5x + 8y + 9z = 6$$

عدد المعادلات لا يقل
عن المجاهيل

Matrix of coeff^{ts}

$$\begin{bmatrix} 1 & 2 & 3 \\ 4 & 6 & 7 \\ 5 & 8 & 9 \end{bmatrix}$$

Augmented matrix

$$\begin{bmatrix} 1 & 2 & 3 & 2 \\ 4 & 6 & 7 & 5 \\ 5 & 8 & 9 & 6 \end{bmatrix}$$

In order to have a unique sol$\underline{n}$ $r = r' = n$
i.e $r = r' = 3$ then the det$\underline{t}$.

$$\begin{vmatrix} 1 & 2 & 3 \\ 4 & 6 & 7 \\ 5 & 8 & 9 \end{vmatrix} \neq 0 \text{ should not be equal zero.}$$

Since $\text{Det}\underline{t} = (54 - 56) + 2(35 - 36) + 3(32 - 30)$
$= -2 - 2 + 6 \neq 0$

Now solve the eq$\underline{ns}$ we have first to write the eq$\underline{ns}$ in the matrix form:-

$$\begin{bmatrix} 1 & 2 & 3 \\ 4 & 6 & 7 \\ 5 & 8 & 9 \end{bmatrix} \begin{bmatrix} x \\ y \\ z \end{bmatrix} = \begin{bmatrix} 2 \\ 5 \\ 6 \end{bmatrix}$$

$$\therefore \begin{bmatrix} x \\ y \\ z \end{bmatrix} = \begin{bmatrix} 1 & 2 & 3 \\ 4 & 6 & 7 \\ 5 & 8 & 9 \end{bmatrix}^{-1} \begin{bmatrix} 2 \\ 5 \\ 6 \end{bmatrix}$$

عن طريقة

$\frac{\text{cofactor}}{\text{det}}$ نحسب / نحسب

<u>Now</u>

$$\therefore \begin{bmatrix} 1 & 2 & 3 \\ 4 & 6 & 7 \\ 5 & 8 & 9 \end{bmatrix}^{-1} = \begin{bmatrix} \frac{-2}{2} & -\frac{1}{2} & \frac{2}{2} \\ \frac{6}{2} & \frac{-6}{2} & -\frac{2}{2} \\ \frac{-4}{2} & \frac{5}{2} & \frac{-2}{2} \end{bmatrix}$$

$$\begin{matrix} + & - & + & - \\ - & + & - & + \\ + & - & + & - \\ - & + & - & + \end{matrix}$$

wrong

very wrong

Right

$$\therefore \begin{bmatrix} x \\ y \\ z \end{bmatrix} = \begin{bmatrix} -\frac{2}{2} & -\frac{1}{2} & \frac{2}{2} \\ \frac{6}{2} & -\frac{6}{2} & \frac{2}{2} \\ -\frac{4}{2} & \frac{5}{2} & -\frac{2}{2} \end{bmatrix} \begin{bmatrix} 2 \\ 5 \\ 6 \end{bmatrix}$$

$$\begin{bmatrix} -\frac{2}{2} & \frac{6}{2} & -\frac{4}{2} \\ -\frac{1}{2} & \frac{-6}{2} & \frac{5}{2} \\ \frac{2}{2} & \frac{2}{2} & -\frac{2}{2} \end{bmatrix}$$

$$x = \frac{-4 + 36 - 24}{2} = \frac{2}{2} = 1$$

$$y = \frac{-2 - 30 + 3}{2} = \frac{-2}{2} = -1$$

$$z = \frac{4 + 10 - 12}{2} = \frac{2}{2} = 1$$

ex 3- examine the following eq$\underline{ns}$. { if there is sol$\underline{n}$ or more than sol$\underline{s}$ or no sol$\underline{n}$

$$2x+y-z=1$$
$$x-2y+z=3$$
$$4x-3y+z=5$$

Matrix of coeff$\underline{ts}$

$$\begin{bmatrix} 2 & 1 & -1 \\ 1 & -2 & +1 \\ 4 & -3 & 1 \end{bmatrix}$$

Augmented matrix

$$\begin{bmatrix} 2 & 1 & -1 & 1 \\ 1 & -2 & 1 & 3 \\ 4 & -3 & 1 & 5 \end{bmatrix}$$

Be care when calculate the rank here the det$\underline{t}$ of the matrix of the coeff$\underline{ts}$ is zero by actual calculat$\underline{ns}$

$$= 2(-2+3) + (4-1) + (-8+3) = 2+3-5 = 0$$

the 2$\underline{nd}$ order determinant. $\begin{vmatrix} 2 & 1 \\ 1 & -2 \end{vmatrix} \neq 0$

or $\begin{vmatrix} 1 & -2 \\ 4 & -3 \end{vmatrix} \neq 0$ or $\begin{vmatrix} 1 & -1 \\ 2 & +1 \end{vmatrix} \neq 0$ $\therefore$ $r = 2$

there is the 2nd order det$\underline{t}$ is non zero.

in the augmented matrix $\begin{bmatrix} 1 & -1 & 1 \\ -2 & 1 & 3 \\ -3 & 1 & 5 \end{bmatrix}$ its det$\underline{t}$ $\neq 0$

as

$$(5-3)+(-10+9)+(-2+3)$$

$$2-1+1=2 \neq 0$$

$\therefore$ $r' = 3$

and since $r' > r$.

Then the eq$\underline{ns}$ are (Inconsistent.)

Note: the Def^{n} of the Rank of a matrix.

A matrix is said to be of rank (r). if there exists at least. one - non zero - determinant of order r while all dett^{s} of order higher than (r) are all zero.

All. determinants higher than (r).

i.e. in the last example we may not say that the augmented matrix of order (rank) 2 as there is exists a dett of order 3 $\neq 0$.

or.

Perhaps the matrix of rank (r) and there exists a dett of order $r = 0$.

But it is necessary if the matrix of rank r all the detts of order higher than r must = zero

10. m Linear Homogeneous Equations in n Unknowns

The preceding discussion has dealt with non homogeneous linear equations. In case the hi are all zero, the system becomes the set of homogeneous equations

$$a_{11} x_1 + a_{12} x_2 + \cdots + a_{1n} x_n = 0,$$
$$a_{21} x_1 + a_{22} x_2 + \cdots + a_{2n} x_n = 0,$$
$$\cdots\cdots\cdots\cdots,$$
$$a_{m1} x_1 + a_{m2} x_2 + \cdots + a_{mn} x_n = 0$$

Obviously $x_1 = x_2 \ldots = x_n = 0$ is a solution of the system It. may happen that there are other solutions. If $\alpha_1, \alpha_2, \ldots, \alpha_n$ is a solution, it is evident that $k\alpha_1, k\alpha_2, \ldots, k\alpha_n$ where k is an arbitrary constant will also be a solution .

The condition for solutions different from the $x_1 = x_2 = \ldots = x_n = 0$ solution will be stated as follows.

The system will have a solution different from the trivial solution $x_1 = x_2 = \ldots = x_n = 0$ if the rank of the matrix of the coefficients is less than n.

Example (12)
Consider the system

$3x - 2y + 4z = 0,$
$2x + 3y - 2z = 0,$
$4x + 5y + 2z = 0$

$$\text{Let } A = \begin{bmatrix} 3 & -2 & 4 \\ 2 & 3 & -2 \\ 4 & 5 & 2 \end{bmatrix}$$

$|A| = 64 \neq 0$

Therefore $x = 0$, $y = 0$, $z = 0$ is the only solution.

Example (13)
Consider

$4x - 3y = 0,$
$2x + 5y = 0,$

$3x - 2y = 0$

The matrix of the coefficients is of rank 2. since the number of the unknowns is 2, $x = 0$, $y = 0$ is the only solution of the given equations.

Example

Example (14)
Consider

$2x - y + 3z = 0,$
$3x + 2y + z = 0,$
$x + 3y - 2z = 0$
$5x + y + 4z = 0$

The matrix of the coefficients is

$$\begin{bmatrix} 2 & -1 & 3 \\ 3 & 2 & 1 \\ 1 & 3 & -2 \\ 5 & 1 & 4 \end{bmatrix}$$

which is of rank 2. Since the number of unknowns is 3, the system has solutions other than $x = 0$, $y = 0$, $z = 0$ Let $z = k$ and solve any two of the equations for x and y. If the first two are chosen, $x = -k$ and $y = k$. By substitution, it is easily verified that $x = -k$, $y = k$, $z = k$ satisfies all the four equations for any choice of k.

Example (15)
Consider

$3x - 2y - 2z = 0$
$x + 2y - 6z = 0$

The matrix of the coefficients is

$$\begin{bmatrix} 3 & -2 & -2 \\ 1 & 2 & -6 \end{bmatrix}$$

which is of rank 2. Since the number of unknowns is greater than the number of equations; there exist other solutions. Let $z = k$ and solve the two equations for x and y we obtain $x = 2k$, $y = 2k$ Thus $x = 2k$, $y = 2k$, $z = k$, is a solution for any choice of k.

Example (16)
Consider

$5x + 2y — 3z = 0$,
$x + y - z = 0$,
$2x + 5y - 4z = 0$

$$\text{Let } A = \begin{bmatrix} 5 & 2 & -3 \\ 1 & 1 & -1 \\ 2 & 5 & -4 \end{bmatrix}$$

$|A| = 0$

Since the determinant of A is zero, there are solutions different From $x = 0$, $y = 0$, $z = 0$. Let $z = k$ and solve any two of the equations. If the first two are chosen, $x = k / 3$, $y = 2k / 3$, $z = k$. Hence the general solution or the given equations can be written in the form:
$x = \lambda, y = 2\lambda, z = 3\lambda$

11. Linear Transformations

Suppose that the variables x_1, x_2 are replaced by new variables y_1, y_2 such that

$$x_1 = a_{11} y_1 + a_{12} y_2,$$
$$x_2 = a_{21} y_1 + a_{22} y_2$$

This can be written in the form

$$\begin{bmatrix} x_1 \\ x_2 \end{bmatrix} = \begin{bmatrix} a_{11} & a_{12} \\ a_{21} & a_{22} \end{bmatrix} \begin{bmatrix} y_1 \\ y_2 \end{bmatrix}$$

or $X = A Y$

The matrix $\begin{bmatrix} a_{11} & a_{12} \\ a_{21} & a_{22} \end{bmatrix}$ is called the matrix of the transformation. Again, suppose that the variables y_1 and y_2 are replaced by new variables z_1 and z_2 such that

$$y_1 = b_{11} z_1 + b_{12} z_2$$
$$y_2 = b_{21} z_1 + b_{22} z_2$$

i.e

$$\begin{bmatrix} y_1 \\ y_2 \end{bmatrix} = \begin{bmatrix} b_{11} & b_{12} \\ b_{21} & b_{22} \end{bmatrix} \begin{bmatrix} z_1 \\ z_2 \end{bmatrix}$$

or $Y = B Z$

$$x_1 = a_{11} (b_{11} z_1 + b_{12} z_2) + a_{12} (b_{21} z_1 + b_{22} z_2)$$
$$= (a_{11} b_{11} + a_{12} b_{21}) z_1 + (a_{11} b_{12} + a_{12} b_{22}) z_2$$

$$x_2 = a_{21}(b_{11} z_1 + b_{12} z_2) + a_{22}(b_{21} z_1 + b_{22} z_2)$$

$$= (a_{21} b_{11} + a_{22} b_{21}) z_1 + (a_{21} b_{12} + a_{22} b_{22}) z_2$$

$$\therefore \begin{bmatrix} x_1 \\ x_2 \end{bmatrix} = \begin{bmatrix} a_{11} b_{11} + a_{12} b_{21} & a_{11} b_{12} + a_{12} b_{22} \\ a_{21} b_{11} + a_{22} b_{21} & a_{21} b_{12} + a_{22} b_{22} \end{bmatrix} \begin{bmatrix} z_1 \\ z_2 \end{bmatrix}$$

$$= \begin{bmatrix} a_{11} & a_{12} \\ a_{21} & a_{22} \end{bmatrix} \begin{bmatrix} b_{11} & b_{12} \\ b_{21} & b_{22} \end{bmatrix} \begin{bmatrix} z_1 \\ z_2 \end{bmatrix}$$

Hence we have the following result:

If X =A Y, Y = B Z, then A B Z

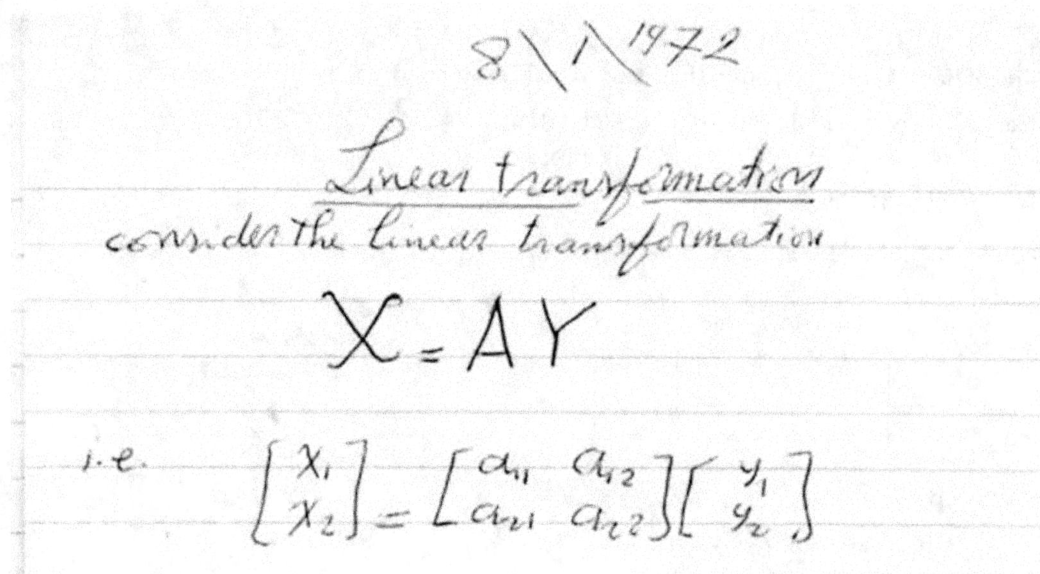

8\1\1972

Linear transformation

consider the linear transformation

$$X = AY$$

i.e. $\begin{bmatrix} x_1 \\ x_2 \end{bmatrix} = \begin{bmatrix} a_{11} & a_{12} \\ a_{21} & a_{22} \end{bmatrix} \begin{bmatrix} y_1 \\ y_2 \end{bmatrix}$

A is called the matrix of a transformation

$\therefore \quad x_1 = a_{11} y_1 + a_{12} y_2$
$x_2 = a_{21} y_1 + a_{22} y_2$ ——— (1)

Now suppose we make a further transformation.

$$Y = BZ$$

i.e

$$\begin{bmatrix} y_1 \\ y_2 \end{bmatrix} = \begin{bmatrix} b_{11} & b_{12} \\ b_{21} & b_{22} \end{bmatrix} \begin{bmatrix} z_1 \\ z_2 \end{bmatrix}$$

$y_1 = b_{11} z_1 + b_{12} z_2$
$y_2 = b_{21} z_1 + b_{22} z_2$ ——— (2)

Substituting from (2) in (1) to eliminate (y)

$x_1 = a_{11} (b_{11} z_1 + b_{12} z_2) + a_{12} (b_{21} z_1 + b_{22} z_2)$
$x_2 = a_{21} (b_{11} z_1 + b_{12} z_2) + a_{22} (b_{21} z_1 + b_{22} z_2)$

$$x_1 = (a_{11}b_{11} + a_{12}b_{21})z_1 + (a_{11}b_{12} + a_{12}b_{22})z_2$$
$$x_2 = (a_{21}b_{11} + a_{22}b_{21})z_1 + (a_{21}b_{12} + a_{22}b_{22})z_2$$

i.e
$$\begin{bmatrix} x_1 \\ x_2 \end{bmatrix} = \begin{bmatrix} a_{11}b_{11} + a_{12}b_{21} & a_{11}b_{12} + a_{12}b_{22} \\ a_{21}b_{11} + a_{22}b_{21} & a_{21}b_{12} + a_{22}b_{22} \end{bmatrix} \begin{bmatrix} z_1 \\ z_2 \end{bmatrix}$$

$$= \begin{bmatrix} a_{11} & a_{12} \\ a_{21} & a_{22} \end{bmatrix} \begin{bmatrix} b_{11} & b_{12} \\ b_{21} & b_{22} \end{bmatrix} \begin{bmatrix} z_1 \\ z_2 \end{bmatrix}$$

$$\therefore X = ABZ$$

Hence we have the following result concerning Linear transformation: mainly

if $X = AY$
& $Y = BZ$
then $X = ABZ$

12. Orthogonal Transformation

In the transformation X = A Y, if the matrix A of the transformation is orthogonal, then the transformation is also called orthogonal. For example,

$$x = X\cos\theta - Y\sin\theta$$

$$y = X\sin\theta + Y\cos\theta$$

the matrix of the transformation $\begin{bmatrix} \cos\theta & -\sin\theta \\ \sin\theta & \cos\theta \end{bmatrix}$

is orthogonal, then the transformation: is also orthogonal (since axes are rectangular).

We shall now prove the following property of orthogonal transformations:

Let X = A Y he an orthogonal transformation, i e.,

A' A = A A' = I, then X' = Y' A'

$\therefore$

$$X'X = Y'A'AY = Y'IY = Y'Y$$

$$\therefore [x_1 \ x_2 \ldots\ldots x_n] \begin{bmatrix} x_1 \\ x_2 \\ \cdot \\ \cdot \\ x_n \end{bmatrix} = [y_1 \ y_2 \ldots\ldots y_n] \begin{bmatrix} y_1 \\ y_2 \\ \cdot \\ \cdot \\ y_n \end{bmatrix}$$

$$\therefore x_1^2 + x_2^2 + \ldots + x_n^2 = y_1^2 + y_2^2 + \ldots\ldots + y_n^2$$

This can also he taken as an alternative definition for an orthogonal transformation. It means that the distance of a point (x_1 , x_2 , . . . , x_n) in space of n dimensions from the origin is an invariant to any rotation of rectangular axes through the origin.

Orthogonal transformation

If $X = AY$

and A is an orthogonal matrix then the transformation is said to be "Orthogonal"

e.g.

$$x = X\cos\theta - Y\sin\theta$$
$$y = X\sin\theta + Y\cos\theta$$

Then i.e: $$\begin{bmatrix} x \\ y \end{bmatrix} = \begin{bmatrix} \cos\theta & -\sin\theta \\ \sin\theta & \cos\theta \end{bmatrix} \begin{bmatrix} X \\ Y \end{bmatrix}$$

This matrix is an orthogonal matrix hence the trans-formation is orthogonal

13. Reciprocal Transformations

If X = AY where A is a non singular square matrix, then $Y = A^{-1}X$

Each transformation is said to be reciprocal to the other.

14. Review on Matrix Definitions

Definations of matrices

1- Unit matrix

$$I_n = \begin{bmatrix} 1 & 0 & 0 \\ 0 & 1 & 0 \\ 0 & 0 & 1 \end{bmatrix}$$

n = number of rows or columns.

---

2. Scalar matrix

$$A = \begin{bmatrix} k & 0 & 0 & 0 \\ 0 & k & 0 & 0 \\ 0 & 0 & k & 0 \\ 0 & 0 & 0 & k \end{bmatrix}$$

Principal diagonal is integers numbers.

3. transposed matrix

$$A = \begin{bmatrix} a_{11} & a_{12} \cdots & a_{1n} \\ a_{21} & a_{22} \cdots & a_{2n} \\ a_{m1} & a_{m2} \cdots & a_{mn} \end{bmatrix}$$

transposed

$$A' = \begin{bmatrix} a_{11} & a_{21} \cdots & a_{m1} \\ a_{12} & a_{22} \cdots & a_{m2} \\ a_{1n} & a_{2n} & a_{mn} \end{bmatrix}$$

rows transposed to be columns

4. Adjugate or adjoint matrix

$$A = \begin{bmatrix} a_{11} & a_{12} \cdots & a_{1n} \\ a_{21} & a_{22} & a_{2n} \\ a_{m1} & a_{m2} & a_{mn} \end{bmatrix}$$

$$(\text{adj } A) = \begin{bmatrix} A_{11} & A_{21} & A_{m1} \\ A_{12} & A_{22} & A_{m2} \\ A_{1n} & A_{2n} & A_{mn} \end{bmatrix}$$

A_{mn} = the determinant

5. cofactor

cofactor $a_{11} = + A_{11}$

" $a_{21} = - A_{21}$

according scheme

$$\begin{matrix} + & - & + & - \\ - & + & - & + \\ + & - & + & - \end{matrix}$$

6. Minor.

minor $a_{11} = A_{11}$

" $a_{21} = A_{21}$

7. Inverse or Reciprocal matrix

$$A^{-1} = \frac{adj(A)}{|A|}$$

$$A^{-1} = \begin{bmatrix} \frac{A_{11}}{|A|} & \frac{A_{21}}{|A|} & \frac{A_{m1}}{|A|} \\ \frac{A_{12}}{|A|} & \frac{A_{22}}{|A|} & \frac{A_{m2}}{|A|} \\ \frac{A_{1n}}{|A|} & \frac{A_{2n}}{|A|} & \frac{A_{mn}}{|A|} \end{bmatrix}$$

8. Orthogonal matrix

if $A A' = I_n$

(matrix) x (transposed of A) = unit matrix I_n

9. Conjugate complex matrix

$$A = \begin{bmatrix} a+ib & c+ih \\ e+if & g+ik \end{bmatrix}$$

$$\text{conjugate } \bar{A} = \begin{bmatrix} a-ib & e-ih \\ e-if & g-ik \end{bmatrix}$$

10- Hermitian matrix

$$\text{if } A = \bar{A}'$$

matrix = transposed of conjugate

11- Skew-Hermitian

$$A = -\bar{A}'$$

12. Latent Roots (characteristic eqn. of a matrix)

if A = matrix I_n = unit matrix

$$|A - \lambda I_n| = 0$$

dett

λ = Latent roots

15. Exercises on matrices

1) If

$$\begin{bmatrix} 2 & 3 \\ 4 & 5 \end{bmatrix} = \begin{bmatrix} 2 & x+y \\ x-y & 5 \end{bmatrix}$$

find x and y

$$\left[\frac{7}{2}, -\frac{1}{2}\right]$$

2) If

$$A = \begin{bmatrix} 1 & 2 \\ 3 & 4 \\ 5 & 6 \end{bmatrix}, B = \begin{bmatrix} 2 & -1 \\ 3 & -2 \\ 0 & 1 \end{bmatrix}, C = \begin{bmatrix} 4 & 2 \\ 1 & 0 \\ 2 & -4 \end{bmatrix}$$

Compute

i) A + B
ii) A + B + C
iii) A - B
iv) B -A
v) (1/2) A – (1/3) C

vi) X, if 3 B + X = 2 C

3) If

$$A = \begin{bmatrix} 2 & 1 & 2 \\ 3 & 5 & 7 \\ 1 & 0 & 1 \end{bmatrix}, \quad B = \begin{bmatrix} -3 & 1 & 0 \\ 6 & 2 & 1 \\ 1 & -1 & 2 \end{bmatrix}$$

find AB and BA

Ans: $AB = \begin{bmatrix} 2 & 2 & 5 \\ 28 & 6 & 19 \\ -2 & 0 & 2 \end{bmatrix}, \quad BA = \begin{bmatrix} -3 & 2 & 1 \\ 19 & 16 & 27 \\ 1 & -4 & -3 \end{bmatrix}$

4) Perform the following matrix multiplications, when possible

a) $\begin{bmatrix} -1 & 2 & 3 \\ 2 & 1 & -2 \end{bmatrix} \begin{bmatrix} 2 \\ -1 \\ 3 \end{bmatrix}$ b) $\begin{bmatrix} 1 & -1 \\ 2 & 1 \\ 4 & 2 \end{bmatrix} [2 \quad 3]$

c) $\begin{bmatrix} 1 & -1 \\ 2 & -1 \\ 4 & 2 \end{bmatrix} \begin{bmatrix} 2 \\ 3 \end{bmatrix}$ d) $\begin{bmatrix} -1 & 2 & 3 \\ 2 & 1 & -2 \end{bmatrix} \begin{bmatrix} 2 \\ 3 \end{bmatrix}$

Ans: a) $\begin{bmatrix} 5 \\ -3 \end{bmatrix}$, b) not compatible, c) $\begin{bmatrix} -1 \\ 1 \\ 14 \end{bmatrix}$,

d) not compatible

5) If

$$A = \begin{bmatrix} 1 & 1 \\ 0 & 1 \end{bmatrix}, \quad B = \begin{bmatrix} 2 & 0 \\ 3 & 1 \end{bmatrix} \text{ and } C = \begin{bmatrix} 1 & 3 \\ 2 & -1 \end{bmatrix}$$

Compute

a) AB
b) [AB] C
c) BC
d) A [B C]
e) A [B + C] and verify that it equals AB + AC
f) [B + C] A and verify that it equals BA + CA

Ans : a) $\begin{bmatrix} 5 & 1 \\ 3 & 1 \end{bmatrix}$ b) $\begin{bmatrix} 7 & 14 \\ 5 & 8 \end{bmatrix}$ c) $\begin{bmatrix} 2 & 6 \\ 5 & 8 \end{bmatrix}$

d) Same as b) e) $\begin{bmatrix} 8 & 3 \\ 5 & 0 \end{bmatrix}$ f) $\begin{bmatrix} 3 & 6 \\ 5 & 5 \end{bmatrix}$

6) If

$$A = \begin{bmatrix} 2 & 3 \\ 4 & 6 \end{bmatrix}, \quad B = \begin{bmatrix} 1 & 2 \\ 1 & -2 \end{bmatrix}, \quad C = \begin{bmatrix} -2 & -4 \\ 3 & 2 \end{bmatrix}$$

verify that A B = A C

7) If

$$A = \begin{bmatrix} 3 & 1 \\ 6 & 2 \end{bmatrix}, \quad B = \begin{bmatrix} -1 & 3 \\ 3 & -9 \end{bmatrix}$$

verify that AB = 0

8) If A^2 means AA show that

$$\begin{bmatrix} 1 & 0 \\ 0 & 1 \end{bmatrix}^2, \begin{bmatrix} 1 & 0 \\ 0 & -1 \end{bmatrix}^2, \begin{bmatrix} -1 & 0 \\ 0 & 1 \end{bmatrix}^2, \begin{bmatrix} -1 & 0 \\ 0 & -1 \end{bmatrix}^2, \begin{bmatrix} 0 & k \\ \frac{1}{k} & 0 \end{bmatrix}^2$$

where k is any non zero number, are all the same.

9) a) If

$$A = \begin{bmatrix} 1 & 1 \\ 0 & 1 \end{bmatrix} \text{ and } B = \begin{bmatrix} 2 & 0 \\ 3 & 1 \end{bmatrix}$$ determine if

AB = BA

b) If

$$C = \begin{bmatrix} 2 & 1 \\ 1 & 1 \end{bmatrix} \text{ and } D = \begin{bmatrix} 2 & -1 \\ -1 & 3 \end{bmatrix}$$

determine whether C D = D C

[Answer: a) no b) yes]

10) Determine, if possible, values of x and y so that if

a) $A = \begin{bmatrix} x & 2 \\ -2 & y \end{bmatrix}$ and $B = \begin{bmatrix} 2 & 1 \\ -1 & 1 \end{bmatrix}$, then $AB = BA$

b) $A = \begin{bmatrix} x & 2 \\ -1 & y \end{bmatrix}$ and $B = \begin{bmatrix} 2 & 1 \\ -1 & 1 \end{bmatrix}$, then $AB = BA$

[Answer a) any x, y =x-2 b) no (x,v)]

11) If

$A = \begin{bmatrix} 2 & -4 \\ -4 & 8 \end{bmatrix}$ and $B = \begin{bmatrix} 4 & 2 \\ 2 & 1 \end{bmatrix}$

show that AD = 0 = BA

12) If

$A = \begin{bmatrix} 1 & 2 \\ 2 & 4 \end{bmatrix}$, $B = \begin{bmatrix} -2 & 4 \\ 1 & -2 \end{bmatrix}$

show that $AB = 0 \neq BA$

13) Find the reciprocal matrix of

$$A = \begin{bmatrix} 1 & 2 & 3 \\ 1 & 0 & 2 \\ 1 & -3 & 0 \end{bmatrix}$$

and verify that $AA^{-1} = A^{-1}A = I$

$$\text{Ans: } \begin{bmatrix} 6 & -9 & 4 \\ 2 & -3 & 1 \\ -3 & 5 & -2 \end{bmatrix}$$

14) If

$$A = \begin{bmatrix} 1 & 0 & 0 & 0 \\ 1 & -1 & 0 & 0 \\ 1 & -2 & 1 & 0 \\ 1 & -3 & 3 & -1 \end{bmatrix}$$

prove that $A^2 = 1$ where 1 is the (4 x 4) unit matrix.

Prove also that if

$P = A M_1 A$ and $Q = A M_2 A$
where M_1 and M_2, are arbitrary diagonal matrices of order (4 x 4),
then PQ =QP

15) Verify that the matrix

$$\begin{bmatrix} 1 & 2-i \\ 3+i & 2 \end{bmatrix}$$

is Hermitian.

Prove that the characteristic roots of a Hermitian matrix are always real

16) Find the symmetric and skew — symmetric parts of the matrix

$$A = \begin{bmatrix} 1 & \frac{3}{2} & -5 \\ \frac{1}{2} & 0 & \frac{3}{4} \\ -1 & \frac{1}{4} & 2 \end{bmatrix}$$

17) show that if

$$A(\theta) = \begin{bmatrix} \cos\theta & \sin\theta \\ -\sin\theta & \cos\theta \end{bmatrix}$$

then $A(\theta)\,A(\phi) = A(\theta+\phi)$ and give a geometrical interpretation of this result.

18) Find the inverse matrix of

$$A = \begin{bmatrix} 1 & 2 & 3 & 4 \\ 0 & 1 & 2 & 3 \\ 0 & 0 & 1 & 2 \\ 0 & 0 & 0 & 1 \end{bmatrix}$$

$$\text{Ans : } A^{-1} = \begin{bmatrix} 1 & -2 & 1 & 0 \\ 0 & 1 & -2 & 1 \\ 0 & 0 & 1 & -2 \\ 0 & 0 & 0 & 1 \end{bmatrix}$$

19) show that the eigenvalues of the matrix

$\begin{bmatrix} 1 & 3 \\ 2 & 2 \end{bmatrix}$ are 4 and -1

20) Find the characteristic equation of the matrix

$$A = \begin{bmatrix} 2 & 3 & 1 \\ 3 & 1 & 2 \\ 1 & 2 & 3 \end{bmatrix}$$

and prove that the matrices

$$B = \begin{bmatrix} 3 & 1 & 2 \\ 1 & 2 & 3 \\ 2 & 3 & 1 \end{bmatrix} \text{ and } C = \begin{bmatrix} 1 & 2 & 3 \\ 2 & 3 & 1 \\ 3 & 1 & 2 \end{bmatrix}$$

has the same characteristic equations as A

$[\ \lambda^3 - 6\lambda^2 - 3\lambda + 18 = 0\ ]$

21) If A is an orthogonal matrix (i.e., AA' = 1), show that its characteristic roots have all absolute value unity. Find these characteristic roots when,

$$A = \begin{bmatrix} 0 & \cos\theta & \sin\theta \\ \cos\phi & -\sin\theta\sin\phi & \cos\theta\sin\phi \\ \sin\phi & \sin\theta\cos\phi & -\cos\theta\cos\phi \end{bmatrix}$$

$[\ \lambda = 1, -\alpha \pm i\sqrt{1-\alpha^2}$ where $\alpha = \cos^2 \tfrac{1}{2}(\theta - \phi)$)

22) If $S = \begin{bmatrix} \frac{1}{\sqrt{2}} & -\frac{1}{\sqrt{2}} \\ \frac{1}{\sqrt{2}} & \frac{1}{\sqrt{2}} \end{bmatrix}$

show that SS'=1 (i.e., that is orthogonal)

Hence show that if $P = \begin{bmatrix} 1 & 3 \\ 3 & 1 \end{bmatrix}$ then $S P S'$ is a diagonal matrix and obtain its eigen values

diagonal matrix and obtain its eigen values

[Answer: 4, -2]

23) If

$$X = \begin{bmatrix} -\frac{1}{2} & -\frac{\sqrt{3}}{2} & 0 \\ -\frac{\sqrt{3}}{2} & \frac{1}{2} & 0 \\ 0 & 0 & 2 \end{bmatrix}$$

and

$$P = \begin{bmatrix} \frac{1}{2} & \frac{\sqrt{3}}{2} & 0 \\ -\frac{\sqrt{3}}{2} & \frac{1}{2} & 0 \\ 0 & 0 & 1 \end{bmatrix}$$

Prove that $P^{-1}XP$ is a diagonal matrix and X satisfies the equation.

$$X^3 - 2X^2 - X + 2I = 0$$

24) If

$$S = \frac{1}{\sqrt{2}} \begin{bmatrix} 1 & -i \\ -i & 1 \end{bmatrix}$$

show that S (S*)'= I . (Matrices satisfying this condition are said to be unitary matrices)

25) solve, the following equations by matrix methods

$$4x - 3y + z = 11, \quad 2x + y - 4z = -1, \quad x + 2y - 2z = 1$$

[Ans : $x = 3$, $y = 1$, $z = 2$]

26) Investigate the following systems and find solutions wherever the system is consistent :

a) $x-2y=3$
$2x+y=1$
$3x-y=4$

b) $2x+y-z=1$
$x-2y+z=3$
$4x-3y+z=5$

c) $3x+2y=4$
$x-3y=1$
$2x+5y=-1$

d) $2x-y+3z=4$
$x+y-3z=-1$
$5x-y+3z=7$

[a) (1, − 1); b) inconsistent ; c) inconsistent , d) (1, 3k − 2, k)]

27) Investigate for consistency, and obtain non-zero solutions when they exist

a) $x+3y-2z=0$
$2x-y+z=0$

b) $x-2y=0$
$3x+y=0$
$2x-y=0$

c) $3x-2y+z=0$
$x+2y-2z=0$
$2x-y+2z=0$

d) $2x-4y+3z=0$
$x+2y-2z=0$
$3x-2y+z=0$

e) $4x-2y+z=0$

$2x-y+3z=0$

$2x-y-2z=0$

$6x-3y+4z=0$

f) $x+2y+2z=0$

$3x-y+z=0$

$2x+3y+2z=0$

$x+4y-2z=0$

[a) $\left(-\frac{k}{7}, 5\frac{k}{7}, k\right)$, b) $(0,0)$,

c) $(0,0,0)$ d) $\left(\frac{k}{4}, \frac{7k}{8}, k\right)$,

e) $(k,2k,0)$, f) $(0,0,0)$]

28) If $A+I_3=\begin{bmatrix} 1 & 3 & 4 \\ -1 & 1 & 3 \\ -2 & -3 & 1 \end{bmatrix}$ evaluate $(A+I_3)(A-I_3)$

29) If $A=\begin{bmatrix} 0 & -\tan\theta \\ \tan\theta & 0 \end{bmatrix}$ verify that

$$I_2+A=\begin{bmatrix} \cos 2\theta & -\sin 2\theta \\ \sin 2\theta & \cos 2\theta \end{bmatrix}(I_2-A)$$

30) If A is a square matrix verify that $A+A'$ is symmetric.

31) Verify That

$$\frac{1}{3}\begin{bmatrix} -1 & 2 & -2 \\ -2 & 1 & 2 \\ 2 & 2 & 1 \end{bmatrix}$$

is an orthogonal matrix.

32) If $I = \begin{bmatrix} 1 & 0 \\ 0 & 1 \end{bmatrix}$, $i = \begin{bmatrix} \delta & 0 \\ 0 & -\delta \end{bmatrix}$,

$$J = \begin{bmatrix} 0 & 1 \\ -1 & 0 \end{bmatrix}, \quad k = \begin{bmatrix} 0 & \delta \\ \delta & 0 \end{bmatrix}$$

where $\delta^2 = -1$, show That

$i^2 = J^2 = k^2 = -I$ and $i\,j = k$.

33) If $l_1^2 + m_1^2 + n_1^2 = l_2^2 + m_2^2 + n_2^2 = l_3^2 + m_3^2 + n_3^2 = 1$
and $l_1 l_2 + m_1 m_2 + n_1 n_2 = 0$, $l_2 l_3 + m_2 m_3 + n_2 n_3 = 0$,
$l_3 l_1 + m_3 m_1 + n_3 n_1 = 0$

prove that $l_1^2 + l_2^2 + l_3^2 = 1$, $m_1 n_1 + m_2 n_2 + m_3 n_3 = 0$, etc.

34) If $A = \begin{bmatrix} 1 & -1 & 1 \\ 2 & -1 & 0 \\ 1 & 0 & 0 \end{bmatrix}$ verify that

$$A^3 = A^2 A = A A^2 = I_3$$

35 Prove that $\begin{bmatrix} \lambda & 1 \\ 0 & \lambda \end{bmatrix}^n = \begin{bmatrix} \lambda^n & n\lambda^{n-1} \\ 0 & \lambda^n \end{bmatrix}$

36) Show that the adjoint of $A = \begin{bmatrix} -4 & -3 & -3 \\ 1 & 0 & 1 \\ 4 & 4 & 3 \end{bmatrix}$

www.ingramcontent.com/pod-product-compliance
Lightning Source LLC
Chambersburg PA
CBHW082035190526
45165CB00021B/3314

* 9 7 8 1 4 6 1 0 8 4 4 2 6 *